The Succulent Manual

A guide to care and repair for all climates

The Succulent Manual

A guide to care and repair for all climates

By Andrea Afra

The Succulent Manual: A guide to care and repair for all climates by Andrea Afra
Published by Sucs for You - Houston, Texas - www.sucsforyou.com

ISBN: 978-0-578-62141-8

This book is dedicated to the plants that have taught me to be patient with myself, my loved ones who have remained a steady source of light and faith, and to the wonderful friends I've made along this succulent journey.

Special thanks to: My mother Tracy Jewell, grandparents Joe and Gaylene Jewell, and my Granny Nadine for instilling a love of gardening in my earliest years; Omar, Lukas, and Khalil Afra—my husband and sons—for loving me and putting up with all these plants; my mother-in-law Amal Afra for sharing every post about succulents I made; my sister-in-law Tina and her family for being so supportive; my sweetest BFFs Maria, Jessica, and Jody who have cheered me on along the way. Y'all are always in my heart!

Preface

I've long since lost track of how many times someone has told me, "You're the reason I got into succulents." I laugh and tell them I gladly take the blame. I also cannot recall all of the times I've been thanked for helping someone save a plant, yet I can't take all of the credit, and I mean it when I say I'm happy to help but you're the one who cared enough to seek out a solution and act on it.

'The Succulent Manual: A guide to care and repair for all climates' sprouted from the exchanges I've had with other plant-lovers online. It began with an Instagram account called 'Sucs for You!' where I simply posted photos of my growing succulent collection along with care tips, identifications, and challenges I encountered in keeping them alive in the humid heat of Houston, Texas. It wasn't long before others were reaching out for help with identifications of their plants and advice on problems they faced. They asked the same questions I had as a newcomer to succulents, and while there are plenty of resources available online and elsewhere, the importance of climate is rarely factored into their insights. My first question to anyone who asked me for help quickly became, "Where are you located?" And you, my friends, are worldwide!

Since I have a background in web design, I built a site dedicated to succulent advice, SucsForYou.com, and then launched a Youtube channel so I could provide visual demonstrations. I knew my plan was working when people who I'd helped wrote in with positive updates, thrilled to report how well their succulents were doing after they had acted on my advice. It was your feedback and comments expressing gratitude and newfound courage to keeping your plants alive and thriving which bolstered my own courage to begin writing this book.

I wanted to be sure I answered as many questions as possible, from every region and season, so I spent well over two years honing the manuscript while fielding inquiries from concerned succulent owners around the world.

People find true joy and solace in their plants. It feels really, really good to know I've helped someone save their special succulent or feel confident about trying again if one dies. By listening and learning along with thousands of others, I found myself uniquely positioned to help those who are in climates very unlike my own. I became a virtual hub of succulent knowledge to be passed along and decided to package it all up as best as I could in this book.

After reading The Succulent Manual you'll have a solid foundation of the essentials to keeping your succulents happy and multiplying, but the learning never ends when it comes to nature.

Some may think we're crazy for being so passionate about these plants, but I think the real draw is how being part of this global community of succulent lovers really makes one feel like they've found their tribe, which is so important to feeling connected, accepted, and whole. And if you haven't joined us yet, we're waiting for you—you'll find us in the garden area (even if we just went to the store for light bulbs) and of course under hashtag 'succulents!'

Cut flowers die, succulents multiply—
Andrea Afra

Contents

Introduction to The Succulent Manual

Millions of people around the world have found themselves oddly in love with succulents. This attraction tends to start with learning to care for one plant and grows to fill a window sill, then a shelf, and eventually the floor or dining table until we buy more shelves. Does this sound familiar to you? Can you still count your succulents on one hand or does your collection take up all the sunny real estate in and around your home? No matter how many succulents we already have, there's always room for just one more, right?

Maybe you don't have many succulents yet, if any. You worry you lack the green thumb or the climate to keep them alive. Or maybe you've lost confidence after watching them die under your care in the past, and out of mercy for any future victims, you stick to cut flowers because they're already dying when you bring them home. I say give this book a read, then give succulents a chance (or another chance.)

With a bit of insight in how to better understand their needs, you too can experience the joy of keeping these plants. A lot of them. I'm literally up to my neck and surrounded by what some may call 'too-many-succulents.' This is a phrase you'll just need to learn to shrug off because the only folks who utter this nonsense typically don't have any succulents. The solution is to give them a plant or two and wait. Before long, they'll realize they also love them and how silly they were to ever think one could have too many.

The chapters in The Succulent Manual contain detailed sections on the most important topics.

Basic Tips: First, you'll learn about the basic care tips and growing cycles so you have a solid understanding of how to keep succulents alive and happy. This is vital to being able to troubleshoot any problems you may encounter.

Make More Sucs: There are many succulents that can provide a lifetime supply of plants for you and others from just one plant! You'll learn about propagation by leaves, division, cuttings, and seeds.

Succulent SOS: Stretched plants, leaf-drop, pests...the 'Symptoms' section will help you identify problems with your succulents while the following section 'Take Action' provides solutions and repair tips.

Regional Tips: Your climate has a major influence on the obstacles you may run into. This chapter covers the various regional and seasonal challenges whether you're growing outside, indoors, or both.

The Succulent Manual also includes identification tips, genus care guidelines, instructions on building a garden bed for in-ground succulents, fun projects and tasks like cleaning your succulents and drilling drainage holes in containers, a buying guide for plants and supplies, a Plant Journal, and other useful advice.

There's something admirable about a succulent's ability to thrive without our interference and to bounce back from certain death when we act to save it. It makes us feel like we're doing something right, though it's usually just the plant's tenacious nature to survive kicking in. But wait. We must be doing something right if a succulent is strong enough to keep growing.

These insights will help you learn what your succulents are telling you, whether it's:

"Stop watering me!"
"I need more sun!"
"That's too much sun!"
"I've got BUGS!"

Or even...

"You're doing great—I'm perfectly happy!"

There is a very special reward waiting for us in our pots and gardens. It's a contentment not found in many places, a quiet appreciation of our relationship with nature and our influence on each other. For some people, it's as good for the soul as therapy or church. And the satisfaction of seeing a good amount of dirt under our fingernails can't be bought.

Faucaria tigrina and friends

Echeveria 'Tsunami' flowers

Chapter 1
BASIC TIPS

"Listen patiently, quietly and reverently to the lessons, one by one, which Mother Nature has to teach, shedding light on that which was before a mystery, so that all who will, may see and know. She conveys her truths only to those who are passive and receptive. Accepting these truths as suggested, wherever they may lead, then we have the whole universe in harmony with us..."

—Luther Burbank, American botanist and horticulturist, from his 1895 speech for the American Pomological Society on "How to Produce New Fruits and Flowers"

Color and Form

It's common to purchase a succulent that catches our eye because of its colors only to find a few weeks later it has faded from a bright pink or purple to mild green and maybe even changed shape. This is usually because your new plant came from a greenhouse in a region unlike yours where light and temperature are highly controlled to bring out the best features in succulents. Unless you can duplicate these conditions, change is inevitable.

However, when the right influences are aligned their colors and true forms return. After a stretch of several bright, dry days with cool weather, succulents really begin to shine. The combination of dry soil, bright light, and temperatures between 50°F and 70°F (10°–20°C) are ideal for bringing out their 'stressed' colors and transforms many succulents into what I call their final form. Euphorbia tirucalli 'Sticks on Fire' will glow with the most brilliant orange and fuchsia, and Kalanchoe luciae 'Flapjacks' reveal their gorgeous crimson potential. Some varieties can produce a rainbow of colors throughout the year in the right conditions. And it doesn't have to be cold for succulents to remain colorful as long as they get bright light and low water, though if it's also really hot, they may suffer from too much stress and go dormant or die. Decreasing the light while increasing water can unstress a succulent in under a week.

This book will help you understand how much control you have over those color-inducing influences based on where and how you're growing succulents. Whether in a window, on your porch or planted in a garden, climate plays the biggest role in how committed with time and money you'd need to be to keep your plants in their final form year-round. Personally, that's not my goal because it would be quite a bit of money and work. I know because I tried.

Echeveria 'Tippy'
displaying bright red leaf tips

While I do love my grow lights, most of my favorite succulents are varieties that don't rely on cool, dry temperatures to look beautiful. Some aren't even that finicky about a few consecutive days of overcast skies, and that only makes me love them even more.

Farina

Many succulents have a natural protective coating called 'farina.' This epicuticular (outer layer) waxy powder functions as a water repelling sunblock against UV rays. Like a enchanted cloak, farina enrobes the plant with a transparent matte-white finish, converting ho-hum leaf colors into ethereal hues rarely seen on our planet, let alone from a plant. The terms 'farinose,' 'pruinose,' and 'glaucous' are all used to describe this milky white layer, but each are a bit different *(see 'Knowledge Bank—Glossary').*

Farina smudged on Senecio mandraliscae 'Blue Chalksticks'

Those new to succulents may mistake the farina for powdery mildew or some other unwelcome residue and try to wipe it off. Accidentally smudging the coating is easy and common. It's not a big deal when new leaves will eventually replace those that get smeared but there are succulents like columnar cacti which don't have that luxury. A decade of growth can be aesthetically ruined with one loving caress. So handle your plants by the roots as often as possible if they're a farinose variety or cup them from the bottom like a little bird to avoid touching the visible tops of the leaves. You may have to warn others not to touch these succulents. Explain why then guide them to some fuzzy Kalanchoes or a bumpy Gasteria, and tell them they're welcome to pet those instead.

Color Codes

The colors of a succulent can tell us a lot about what it needs to thrive. The greener a plant is, the more chlorophyll it contains. This enables it to photosynthesize with more gusto than a plant with lighter hues such as pastel pink, pale green, or variegation. When skies are gray for a longer spell than usual, succulents with green leaves won't stretch as quickly as others.

The easiest varieties to start with and keep looking tip-top are non-rosette forms with darker green coloring. They're also ideal for filling those part sun spots. There are thousands of really amazing green varieties and I listed some I really love in *'Chapter 9—Buying Guide.'* They include Sansevierias, Haworthias, Kalanchoes, Euphorbias, Aloes, and cacti. All of them flower and come in an endless supply of shapes and unique markings.

In other words, you're going to need *a lot* of pots.

Light

I come across a lot of photos from proud new succulent owners that make me think, "Uh oh, that's gonna die," but only if some minor yet urgent changes aren't made. Deadlocked with overwatering, insufficient light is the most common challenge among growers to keeping succulents healthy and in good form. We'll cover grow lights after discussing natural light.

Many plants are heliotropic, meaning 'sun turning' or facing. In the absence of ample light, they literally reach for the sun, growing taller or longer with increasingly weaker stems. This stretching is called etiolation. The most obvious example of light deficiency is seen in rosette-forming varieties. The first sign of etiolation in a rosette shows when the top layer of leaves start to turn towards the light. Rosettes are meant to grow in compact patterns with very little space between each row of leaves. Without enough light, the stem growth surpasses that of the leaves which causes them to grow in spaced further apart. When the distance between leaf rows is too great, rosettes can become unrecognizable.

After several days of overcast weather these *Graptopetalum paraguayense* 'Ghost Plants' began reaching for the light.

Even with the proper amount of light, it's sometimes necessary to intervene a tad to keep rosettes looking rosy. I have a lovely Echeveria 'Lola' that I'll use as an example. Looking directly down at her and from the side, I can see when she has started to tilt her head towards the sun ever so slightly. I then simply turn her pot 180° so her tilt is facing away from the sun. I check again every week or so and repeat as necessary.

The reward for my minimal effort is a fascinating mandala-shaped Lola who turns all the heads of those who walk by.

It's harder to identify etiolation in succulents that aren't such sticklers about

Echeveria 'Lola'

compact growth, such as Aloes, Jades, and Haworthias. As with Echeverias, new growth is often paler than the old, but if it remains lighter than the rest of the leaves it could be a sign more light is wanted. And if you see leaves spreading wider than their normal form, they may be revealing more surface area in an attempt to absorb more light. Etiolated succulents don't always die, but they certainly don't thrive. They're hungry for the light they need to photosynthesize and make food, which means they're more likely to succumb to other deficiencies.

There are some varieties like Sedeveria 'Starburst' that naturally grow long and somewhat spindly stems. These look really great as hanging plants or spilling over the side of a pot.

One of the neatest things about succulents is their propensity to generate new plants from just about any part of the main plant. Not all succulents share this feature but many do and we can use it to our advantage when repairing a stretched plant if we don't fix the lighting problem in time. We'll cover that in *'Chapter 3—Succulent SOS: Take Action - Etiolation Repair,'* but for now we're going to talk about preventing etiolation by finding the right 'dosage' of light.

Determining the differences in full sun, shade, and everything in between can be a challenge, especially since the sun's path changes throughout the year. As sunlight is a dynamic element, we need to observe how it moves across our homes and yards to see what we're working with light-wise. Let's settle on a basic key for talking about sun levels in this discussion. The required

These Sedeveria 'Starburst' grew quite tall before finally cascading over the pot.

Outdoor Light Key

Full sun: Six or more hours of unfiltered/direct sunlight per day. Coupled with high temperatures, many softer succulents will be at risk for sun damage and overheating after more than a few hours of direct sun.

Bright indirect/filtered light: The ideal middle ground for most sun-loving succulents and cacti; at least 6 hours a day. The terms 'bright indirect light' and 'bright filtered light' refer to sunlight that is tempered by the sun's position in the sky, or by items like leafy branches, sheer curtains, and other plants. For instance, I've placed a lawn chair in front of some Echeverias that were getting too much direct sun in my garden. They still received bright indirect light and it was an easy fix that bought me more time to relocate them.

Part sun: Four to six hours of bright sunlight per day, towards the high end if filtered or indirect.

Part shade: Two to four hours of filtered/indirect sunlight per day.

Shade: Less than four hours of unfiltered sunlight per day.

1. **Full/Direct light**
2. **Filtered light**
3. **Indirect light**

sunlight hours don't need to be consecutive. Three hours of morning sun and another five in the afternoon qualify as eight hours of sunlight.

To better understand these light definitions, imagine we're sitting on my west-facing back porch in Houston, Texas, sipping on some iced sweet tea (with lime for me please!) while looking out at the yard on a sunny spring day. Noon is about half an hour away. To the far left, there's a huge old pecan tree with branches reaching across most of the yard that have finally started filling out with new leaves. As it is still spring, the sun's path is closer to the south because Earth's axis is tilted, and even at high noon, it won't be directly overhead until mid-summer. The front door faces east, and as the sun rises, the intensity of the light increases where we are sitting and on the plants surrounding us. At noon, when the roof blocks the sun from directly hitting us, this is 'bright indirect light.' The sky is reflecting the light while the sun is still out of view. Around 2pm the sun will be visible to our eyes, but the pecan leaves filter it partly until sunset, so that's 'bright filtered light.'

Now let's fast forward to mid-summer. The trees have filled out their shady foliage and filter more of the light. However, the sun's path is higher this time of year. It rises earlier so it's brighter outside sooner, and the daylight hours are longer. That means that even though the trees are filtering more light, the extended length of daylight compensates for it being a bit shadier.

And of course all of this depends on where on Earth you live. Those closest to the equator will experience about 12 hours of daylight year round. Those furthest from the equator will have shorter winter daylight hours and much longer summer daylight hours. For example, Fairbanks, Alaska has about 4 hours of daylight during December, but upwards of 21 hours in June. Compare that to Houston with about 10 hours of December daylight and 14 during June.

So as the seasons change along with the sun's path, and as the trees drop and regrow their leaves, your plants may need to be moved several times annually to satisfy their light requirements. I think of it as learning a new dance—a year-long dance with a lot of different partners.

To reiterate, filtered and indirect light simply refer to what is happening between the sun and your plants. In hot temperatures, it's important to find a way to filter or block the sun from directly pounding succulents that can't tolerate extreme heat coupled with bright direct sun.

When bringing home new plants that require bright indirect light, especially if you have them shipped in a box to your door, and doubly especially during a hot summer, it's very important not to overexpose them to the sun right away. Acclimate them to more light by potting them up and placing them in part sun for a few days, then begin moving them near their final home *(see 'Chapter 3–Succulent SOS: Take Action – Increase Light').*

Grow Lights

I know that many of you are dealing with a lack of natural light *(see 'Chapter 4– Regional Tips: Lack of Sunlight').* You have to keep your succulents indoors due to the weather, or you don't have a bright enough spot outdoors. Some regions are notoriously overcast seasonally or year-round. I'm right there with you. While I do have a lot of yard space with bright sun, only my porch is protected from rain, and most of it gets direct sun during the hottest time of the year while not getting enough from winter to late spring.

I decided to get grow lights when I started seeing my succulents lose their form from too many consecutive days of cloudy skies. I also wanted to start some seeds and couldn't guarantee they'd get enough light without an artificial backup. At first I kept my setup in a shed, but it was too hot in the summer. The amount of fans I would need to cool it down would probably trip the breaker. Since I didn't have a place to put them in my house, I set them up on my covered porch.

I have a 4x2-ft (1.2 x .6 m) fixture with six T5 6400K fluorescent bulbs hanging from a stand atop a 6-ft (1.8 m) folding table. The fixture has an option to turn on just two or 4 bulbs but I found 2 bulbs gets the job done. Ideally the light will provide at least 2,000 lumens per square foot. I can fit about 70 various succulents under the lights and I keep them on for about 14 hours a day all year but you can experiment with mimicking the season's natural daylight hours. When it gets really hot and humid here I set a small fan on their table. In the winter, the lights raise the temperature on the table by at least 5 degrees.

I built the stand with plywood and hung the lights with the included adjustable clips. The fixture cost about $120, the table $40, and the stand only about $10 in wood and screws, but you can find decent stands starting at $30 or just use strong ceiling hooks like those used for hanging heavy planters or bicycles. So for less than $200 you can have a reliable light source at the flip of a switch. Considering how much you may spend on succulents over the next few years, it will probably pay for itself in plants saved. You'll also get to enjoy gardening day or night in any weather.

Light Distance from Plants

It's common to hear that you should keep the lights very close to your succulents—within 6" (15 cm) from their tops. I tried this at first, and they didn't like it. Some stopped growing and others were scorched. Also, my plants are of varying heights so I tried using pots and other objects to raise them all to equal levels which quickly became a tedious nuisance. I lifted the lights to about 12" (30 cm) above the shortest succulents and 6–7" (15–18 cm) above the taller pots, and they're much happier with this distance.

Light Colors

Light color is measured in Kelvin (K) and most succulents want at least 5000–7000K. As I mentioned, mine are 6400K. White light is produced by combining all the colors in the visible spectrum in equal proportions. Fluorescent tubes actually produce ultraviolet light, but add an inner coating of phosphor and they emit light that appears white. Most LED grow lights use blue diodes coated with a yellow phosphor to produce light that looks white. You may have seen some grow lights that look pink or purple. These are LEDs made of multiple red and blue diodes.

Each color elicits a different response in plants. Blue light assists vegetative growth, while red helps with flowering. Green and yellow aren't as readily absorbed, but they do serve their functions in the overall growth of plants. If you've ever been in a room with colored lights, you might have found it too unpleasant to stay there for long. They strain my eyes and make everything look monochromatic and surreal. I like fluorescent white lights because they look more natural and offer the full color spectrum plants need to grow. Another bonus is they also help perk me up when the days are gray.

These basic facts will help you get started with grow lights, but with all the fascinating and complex information on how color affects plant growth, it's definitely worth researching further if you're interested.

Beauty is in the eye...

It's obviously important to understand what your plant is telling you about its light situation, but how do we know what it wants if we've never owned a particular variety? With an internet search for photos of the same plant kept by seasoned growers, we can see what they look like under ideal circumstances and attempt to get the same results. Sometimes this is difficult or impossible if our climate isn't like theirs, but we can still aim to mimic the primary requirements and get very close. For help finding names, see '*Chapter 5–Identification.*'

Your Texas-grown Echeveria may be a bit taller or not hold its color as long as one grown in California, but if it's healthy and you do your best to give it what it needs, it will be just as beautiful and delightful—even more so because it's yours, you love it, and you know it loves you right back.

Watering

Several factors play a role in deciding when to water, and more importantly, when not to water your succulents. This is why it's impossible to give a precise watering schedule for every plant and every gardener. The most commonly repeated basic guideline is to allow the soil around the roots to dry out before watering again, but that's way too basic for succulents. The amount of time it takes for the soil to dry out is crucial to the survival of the plant. If a week has passed and the soil isn't dry to the touch around its roots, you run the risk of rot and leaves splitting from engorgement. Edema (or oedema) occurs when the cells burst due to excess water and looks like small blisters, scars, or scale pests.

It's not necessary to flood the pot when watering—that is, to water until it runs through the soil and out the drainage holes—particularly if you live in a humid climate. Succulents don't uptake water through their roots like straws, rather they absorb moisture from the air around their roots. If the soil is too wet for too long, they suffocate. Drying time is determined by the composition of the potting mix, the container it's in, and the climate—including humidity, temperature, and light. By using well-draining soil in pots with ample drainage holes and watching the forecast, you can avoid one of the top causes of succulent death—too much water.

Following is a list of things to consider before wetting your plants:

–*Do you still remember the last time you watered a plant?* Don't water it. (I'm joking, kinda. But if you do remember, you're too organized or you need more plants. *Again, I'm joking. Kinda...*)

–*Is the soil still moist around the roots?* Don't water it. Use your finger or a small wooden stick to carefully check. Like poking a cake with a toothpick to see if it's cooked through, moist soil will cling to the stick, telling you to wait a couple more days to check again. Most succulents can usually stay dry well over a week without issue if it's not too hot outside. Letting the soil remain dry for three days or more also helps prevent mold and pests like fungus gnats from invading your pots.

–*Does the forecast predict overcast weather/rain/humidity that will prevent timely drying of the soil?* Don't water it. Wait until the skies will be clear again. This will also help slow any etiolation due to low light conditions.

–*Is it a beautiful sunny day and you feel the urge to sprinkle some love on your plants with a watering can?* Sorry. Don't water them. Droplets that remain on the leaves and stems can magnify the sun and heat, leaving unsightly burn marks. Wait until closer to sunset to avoid sunburn.

—Did you just get some new cuttings or twist off some leaves to propagate? Don't water them. Wait until any broken or cut ends are callused over and dry to sight and touch before introducing moisture or water will get inside their cells, potentially causing rot. Remember, without roots, they have no way to uptake the water.

—Did you just repot a plant? Don't water it. Broken roots are inevitable and as I mentioned above, if water gets inside of them, they can rot. Let them rest in their new pot and heal for two or three days before watering. It's best to repot in dry soil. Give cacti even longer—up to two weeks as they're more easily shocked by new situations.

Most water-based issues are repairable, and we'll discuss them in 'Chapter 3— Succulent SOS.'

I'm guessing you probably want to know when you should water your succulents. Assuming your soil is drying fast enough and the plant isn't in dormancy (we'll get to that too), you can water about once a week. That's a very basic but important guideline to follow for most succulents. Cacti and others like Lithops are special snowflakes and different rules apply to each. We'll discuss them more in 'Chapter 6—Genus Tips.'

Finally, observe your plant the day after watering to see and feel how it looks when hydrated. Then over the next several days watch for how it shows you when it's ready for a drink. It could be in dry soil a week or more before it even begins to show signs of thirst. You can take photos to compare a recently watered succulent with one that has been dry for a few days. Seeing them side by side will really help you notice the changes.

Soil and Fertilizer

So what's the dirt on succulent potting soil? There are usually one or two pre-bagged options for sale at lawn and home improvement stores marketed under 'succulent and cactus potting mix.' In drier climates you might get away with using this mix without any modifications. However, many gardeners find it too brown for their succulents, meaning it has too much organic soil and not enough grit and drainage. Plus, it's quite pricey because it's a specialty product sold in smaller bags, and the markup is ridiculous unless you really only need to pot a few containers. I mathed it out and the premixes can cost about seven times more than regular garden soil. Finally, a lot of the premixes are amended with fertilizer which is not ideal for succulents and cacti. The wrong nutrients at the wrong time can cause problems like disproportionate leaf-growth, weak stems, and other imbalances in these slow but steady growers. We'll talk about feeding your succulents in a moment, but first let's mix up a big batch of our own fast draining soil.

In order to recreate a good home for these minimalist survivors, we need to look at the soil conditions of their natural habitats. They've adapted to survive in extreme heat and long droughts. If they're not just clinging by their roots from the face of a cliff, they're seeking shade beneath a tree, behind another plant or wedged tightly under an ancient rock. The earth they grow from is hard, rocky, and typically dry. Nutrition is acquired from rain, rocks, runoff water, and photosynthesis.

Now think about all of that while looking at a bag of garden soil. It's rich, dark brown, habitually damp, and sprinkled with fertilizer—basically the exact opposite of the environment described above. To have more control of the soil contents, it's better to start with topsoil. Not only is it cheaper, topsoil is unamended with goodies like time-released plant food because it's not used to produce big juicy tomatoes and peppers but rather to fill holes and level the ground. There may be some larger pieces of bark to remove or chunks to break up, but at $2 or $3 for a 40 pound bag, the cost offsets the little time it takes to sort out any debris you don't want in your pots.

Drainage Materials

Perlite: Moving forward, we're going to need to add some drainage materials to our mix. Perlite tends to be the easiest to find on the shelves and is fine to start with, but I strongly urge you to go the extra mile and seek out some other options as well since perlite retains more moisture than the other materials we'll talk about. Perlite is the white, lightweight, fluffy stuff you see in potting soil. Similar to pumice, it's igneous rock that is processed under extreme heat until it expands and 'puffs.'

Perlite

Pumice

Turface

Expanded shale

Pumice: Pumice is also igneous rock, but it's heavier, harder, and doesn't float to the surface of your pots like perlite. Though it can be more difficult to source locally in most states, you can order it online. This may seem like too much trouble but I promise it's well worth the extra effort. I would much rather do away with perlite altogether for several reasons, but I keep it around because of its availability. Pumice is beautiful! Well, by comparison to perlite at least. It looks like natural organic rock, white to gray in color, and doesn't crush to a fine powder between your fingers like perlite. It's also jam-packed full of minerals your succulents crave. As of writing this, I pay just under $30 for 15 pounds of lovely pumice, including shipping. The big bag of perlite I buy sells for about $20. Both serve their purpose in my soil mix, but perlite doesn't double as a pretty top dressing like pumice. 'Top dressing' is the term for a layer of rocks used to cover the exposed soil in a pot. It protects the bottom leaves from touching wet soil and gives your arrangements a professional, finished look.

Turface: Another option is Turface or porous ceramic soil conditioner. While the word 'Turface' is a brand name, you'll see it used as a general term for this crushed clay amendment. You've probably seen it without realizing it could be used in your succulent mix...during baseball games! You know, the reddish dirt neatly raked to form the diamond—that's the stuff. It comes in a few different particle sizes and colors, ranging from gray to rust red, and it's very affordable. I get 40 pounds for $13 from a local garden store called Southwest Fertilizer. I prefer the larger particle option in red over the gray because when I use it as a top dressing, it looks less like kitty litter. A mix of the two colors looks great together.

Expanded shale: Expanded shale is also a good porous medium to add to your potting mix. It's pretty easy to find, costs about the same as Turface, and it too makes a visually pleasant top dressing. I know this sounds like odd advice, but smell your shale before purchasing it. Shale is often fossiliferous, and as petroleum is a fossil fuel, I've heard a few people complain of it having a crude oil odor, though I haven't come across it myself. In fact, go ahead and smell all of your soil materials before bringing them home, especially the topsoil. It should smell earthy and non-offensive. If it stinks of sulfur or ammonia, it's probably too wet and filled with the wrong kind of bacteria, so try another bag or another brand.

Materials to Avoid

Most sand isn't recommended since it compacts too easily and doesn't allow water to drain quickly. This includes beach sand and store bought sand used for playgrounds, ground leveling, and aquariums. You may have heard of sharp sand. It's different than other sand because its edges aren't rounded like beach sand, but coarse which allows for better drainage. It's helpful as a top layer for seedlings but still not ideal for a potting mix. Crushed gravel and pebbles are also not recommended in humid climates since they aren't porous and can trap excess water in the soil. It's better to just use topsoil and a lot of perlite than pay money for something that could do more harm than good in the long run.

Remember to avoid any materials where the grain size is too small or doesn't allow the soil to evaporate fast enough for succulents.

Top Dressings

Avoid using non-porous rocks such as aquarium gravel and river pebbles as these options trap moisture in the pots which can result in root rot. It's fine to sprinkle a sparse layer of pebbles over the surface, but don't lay them on too thick. For additional decoration, I think a few larger stones work very well. Most of us have found some pretty shells or crystals and pocketed them. Now you have a good excuse to keep bringing home your little treasures—plants like accessories too!

If you live in an arid climate you can get away with using less drainage materials than folks in humid climates. And it's not necessary to use multiple materials, but having different particle sizes helps make the soil sturdier for establishing roots.

Fast Draining Soil Recipe

Supplies:
Topsoil
Drainage materials, 1 or more types
2 bins for mixing (large & small)
Shovel

Start with several shovels full of topsoil, then add an equal amount of perlite or the least costly amendment you have on hand.

Tilt the bin side to side to begin stirring things up, then add a few shovels of your other drainage materials. Keep mixing until you see equal parts soil to other materials. Humid climates will need more drainage materials than arid climates.

This bin is going to get pretty heavy, so transfer a few pots worth to the smaller bin and add a few more handfuls of your pricier drainage materials.

That's it! Very easy and more cost effective than buying a mix that you would still need to amend. For more finicky plants like cacti and Lithops, you should start with even less brown soil, especially in humid regions.

Final product: Equal parts topsoil to other drainage materials

Mix Ratios

Organic % to Drainage %

Aloes- 40/60
Cacti- 20/80
Echeverias - 30/70
Euphorbia- 30/70
Haworthia- 30/70
Jades- 40/60
Kalanchoes- 30/70
Mesembs- 10/90
Sanseviera- 40/70

Fertilizer

Succulents kept outdoors and given rainwater may be fine without food, but many still appreciate a nutrient boost during their growing seasons. On the other hand, skip feedings if you're using a potting mix that already contains fertilizer as too much nitrogen and other nutrients can really freak succulents out.

Feeding your succulents is as simple as mixing up a standard balanced fertilizer and diluting it by half or more, but I suggest you buy a liquid fertilizer made specifically for succulents which has less nitrogen than a more general purpose plant food. You can offer most succulents a meal with each watering. Give them the normal amount of water during feedings and avoid splashing the leaves with fertilizer water as it can cause spots and increases the possibility of sun damage. Don't feed them before or during extended periods of overcast weather to avoid abnormal growth from insufficient light. You should also stop feeding no later than a month before their dormant season to allow time for the nutrients to be processed.

Containers

Your. Pots. Need. Drainage. Holes.

As mentioned in the section on watering, the roots of succulents don't operate like your average plant. Instead, they 'breathe' in the moisture from the air around them which is why we work hard to get our soil loosie-goosie. Drainage holes not only allow water to exit the container, but they assist in airflow and keep the soil around the roots from compacting.

I know, I know—you see all those cute glass terrariums and jars being potted, and if others can use them...? No. Well, fine. Go for it if you must, but please plan on them being temporary arrangements. I agree, they are really cute and fun to make, but I and my fellow adamant pro-holers have good reason to withhold our blessing to use any container that will not drain. A lot of tutorials suggest adding a scoop of rocks before adding soil to a holeless pot so any excess water will seep beneath the roots. It will do exactly that, but then what? Without any contact with soil, there's nothing to wick the moisture back to the surface so it can evaporate. Unless you live in an extremely dry climate, the water will just sit there getting nice and grody, creating a lovely environment for mold and bacteria. Terrariums and pots without holes will eventually smother your succulents if they don't die from root rot first. But if you decide to try using a hole-free container, use soil as your first layer to give

Stacked clay pots with drainage holes all the way down...
See Chapter 8 for how to make this mini nesting pot project.

the water a way to travel up and out.

Better yet, add your own holes. A power drill and the right bits are all you need to get the job done. You can drill metal, plastic, clay, glass, and ceramic. I have friends come over under the guise of wanting to hang out, but what's in the tote bag—wine? 'Oh. Nice mug. No, of course I don't mind if you borrow my drill while you're here. Wine would have been nice though.' I'm really joking! I love teaching people how to use a drill, and it's fun to watch a friend's obvious satisfaction when completing their first hole.

Some materials are harder to drill than others. The blue bowl pictured below took a good half hour to drill 3 holes. Remember that non-porous materials like glazed ceramic or metal will need extra drainage holes, especially if you don't live in an arid climate. And deeper pots will require more drainage materials in the soil. For drilling instructions, see *'Chapter 8—Tasks & Projects: Drilling for Drainage.'*

Soup bowl

This mug just had to be potted!

Tea time! Metal kettle with Woolly Senecio for the steam

Flour pot with M. elegans

Pot Material

The material of your pots contributes greatly to the soil's evaporation rate. Unglazed clay pots are ideal for most succulents as they're porous and draw moisture out of the soil, while plastic and other non-porous containers dry out more slowly. When painting clay pots, leave the inside and the bottom unfinished.

Consider adding some extra drainage holes if you find any pots retaining moisture longer than desired.

Pot Size

You may have heard the warning to avoid overpotting your plants. This is referring to using a planter that holds more soil than your succulents require. Larger pots take longer for the soil to dry so make sure your container fits your plant. This is very important to keeping cacti healthy. Adding multiple plants will help the soil dry more quickly.

The size of your container can also influence the size of your plant. I had a bunch of Kalanchoe laetivirens plantlets and potted some in a little vintage dessert bowl, and another one in a drilled cork. The others were turned loose into a big planter and the results were what you'd think. The plants left in small containers stayed very small, while their siblings are now up to my knees and still growing.

Growth and Dormancy

Succulents are slow growers. Some, very slow. If you're used to tropicals and other plants that can double in size in a single season, get ready to learn a lesson in patience. After spending a couple of years with the same variety, you'll become familiar with its tendency to grow more during certain months than others. Those periods of time are called the 'growing seasons' or 'growing periods.' Where spring and fall offer the most temperate weather, those will likely be the growing seasons for succulents in that region. Colder climates will see more new growth in early summer since their spring months can still freeze.

While most plants do have a period of dormancy that usually takes place during the coldest months, if your region doesn't experience long stretches of low temperatures, it's possible your succulents won't go completely dormant. This seems like a good thing until you factor in the decreasing daylight hours of winter and what that can do to certain varieties, especially cacti. If succulents are still actively growing but not getting enough time in the sun, they will become etiolated. Luckily, the cold tends to slow their growth naturally.

During the hottest months, many succulents grow more slowly to conserve energy. Like Aeoniums, some are inherently summer dormant. However, the majority are 'opportunistic' growers, meaning whether kept indoors or out, they continue to grow when given the right temperatures, water, and light. But there are times when you might need to intervene and create your own dormant period for some plants.

For instance, since I keep my plants outdoors, my eye is always on the forecast because I don't want my succulents trying to grow in poor sunlight. So if there's a good chance the weather is going to be cruddy and gray for two or more days in a row later that week, I don't want to encourage them to grow. Instead, I'll wait to water or feed my plants until the report shows clear skies for at least two or three days back to back. I usually don't water prior to leaving town for less than a week. It's kind of like pushing pause on your succulents' growth so you don't have to worry over the weather as much while you're gone.

Overwintering

'Overwintering' plants is a method of slowing growth by watering just enough to keep the roots alive. If kept in a cool location, succulents won't need as much light as they would in the warmth of your heated home. And as mentioned in 'Chapter 1— Basic Tips: Soil and Fertilizer,' stop feeding by mid-summer so they have time burn up that extra fuel. If you live where it freezes, you already know you're going to need to bring your plants indoors. In this situation, it's good to know how to overwinter succulents and keep them happy until you can put them back outside again (see 'Chapter 4–Regional Tips: Indoors').

I overwinter my cacti and other succulents that are deciduous and have a large caudex. Unless it's going to freeze, I leave them on my porch and only water lightly about once every six weeks or less. If the temperatures drop below 40°F (4.5°C) for more than two days, I may try to find a warmer spot for them away from the screens and closer to the wall. And when it does freeze, I try to bring my favorites indoors while piling the rest under the grow light on my porch. Those that don't fit under the lights get swaddled with warm things like frost cloths, blankets, towels, sweaters, beanies, or whatever is convenient.

You can also overwinter your plants in a cold space that stays above freezing, like a basement or garage, but their soil should be fully dry. The chilly temperatures will slow their growth so they need less light. A few hours of indirect sunlight or a grow light will suffice but remember not to water during this time. Use a fan to help with humidity and air circulation if necessary. When it's time for them to go back outside be sure to ease them into the light and don't water them too deeply at first.

Summer Dormant Genera

Adromischus	Haworthia
Aeonium	Kalanchoe
Aloe	Pachyphytum
Anacampseros	Pachyveria
Conophytum	Peperomia
Cotyledon	Portulacaria
Crassula	Sansevieria
Dioscorea	Sedeveria
Fouqueria	Sedum
Gasteria	Senecio
Graptopetalum	Tylecodon
Graptoveria	

Winter Dormant Genera

Adenium	Ibervillea
Agave	Ipomoea
Alluaudia	Jatropha
Bursera	Lithops
Ceropegia	Pachypodium
Cissus	Sinningia
Dorstenia	Stapelia
Echeveria	Tillandsia
Euphorbia	Xerosicyos
Fockea	
Huernia	

Summer Dormancy

While some summer dormant varieties still act like opportunistic growers, others have an 'obligate' dormancy and can die if you encourage growth when they need to rest. I only have a few succulents that are truly summer dormant because my home doesn't get a lot of good light inside where it's nice and cool, and that's what these types of plants really need.

One is an Albuca spiralis that drops its curly leaves by early summer. I put the potted bulb inside by a bright window and water it maybe once or twice in the four months it's kept indoors. When I start to see new growth, I put it back outside and start watering it again.

Aeoniums also tend to be summer dormant. They may drop leaves and tighten up their rosettes. My friend gave me a cutting from her Oakland garden in late spring, and luckily I was able to get it rooting indoors since it would try to go dormant in the heat outside and be less likely to put out any new growth.

A 'keiki' growing on the flower stalk of a Graptoveria 'Fred Ives'

Wake Up!

You've made it through the winter and you can't wait for spring to awaken your plants. This is when you will be rewarded for keeping them alive through the most trying months of heat or cold. I can't help to feel like they're telling me I've done a good job as they proudly display new leaves, new life! Nothing quite compares to the satisfaction of watching the results of your care and patience unfurl in the garden. That is, until you see your first sign of flowers...

Flowers and Seeds

Good news y'all: Succulents are angiosperms (they flower) and the majority of them are polycarpic, meaning they bloom multiple times in their lives. Granted, there are many varieties whose flowers aren't exactly exciting compared to others, but that doesn't detract from the pleasure of watching them grow. A lot of Haworthias will put out flower stalks that are laughably long, especially when you finally see their tiny white blooms. But the nectar of these wee flowers are sweeter than the brightest coral Aloe maculata blooms. I've tasted Echeveria 'Lola' nectar too, and my Haworthia is still sweeter.

Most of my succulents don't bear scented flowers, but the handful that do reveal a range of fragrances from delightful to 'Oh my gosh, why did I just smell that again?' The blooms of my Crassula *'Morgan's Pink'* and Echinopsis subdenudata *'Domino Cactus'* would make the loveliest perfumes, while the flowers of Stapeliads have evolved to attract flies as pollinators and smell like everything from fish food to a small dead rodent. Yet these 'stank-flowers' (not a botanical term, but feel free to use it) are so wild and unearthly in form, you can't pass them up just because they stink. There are plenty of varieties with smaller blooms that you can't smell unless you're really close to the flowers. Just hang them somewhere downwind while they're in bloom if their odor is too strong.

Most succulents tend to flower from spring through fall. Many types require vernalization, which is the need for a period of cold weather to induce blooms. And at first, it can be hard to tell if your plant is growing a bloom or a new plant. The flower stalk usually starts higher up the stem but a few varieties, such as Echeveria prolifica, grow babies on long stems that look very much like a flower stalk.

Flowers are a great help in confirming the identities of those unknown succulents in your collection *(see 'Chapter 5–Identification').* Some varieties create leaves along the stalk that you can propagate, and on rare occasion, particularly with certain Haworthia, a new plantlet will form high up on the stalk. These are called *bulbils* or *keikis*, a Hawaiian word for 'little one.' I've only had this happen twice— once with an Echeveria diffractens and a Graptoveria 'Fred Ives.' They're such a lucky surprise!

Parodia arnostiana flowers, blooming in April

Succulent blooms can last a long time and can put on a colorful display for months. Once the show is over, just trim back the flower stalk at the bottom to remove it, or if you have seen pollinators enjoying the blooms, you may want to wait to see if it goes to seed. Some genera are self-fertile while others require cross-pollination.

There are a few downsides to flowers. They take a lot of energy from the main plant to grow, the nectar can attract sooty mold, and they can be very appealing to aphids and mealybugs. If your succulent seems weak, you can cut the flower stalks off and enjoy them in an empty vase. And if you see mealies on the blooms, trim them off and throw them away...far, far away from your other plants, and make sure the rest of that plant is free from pests *(see 'Chapter 3—Succulent SOS: Symptoms – Signs of Common Pests').*

I encourage you to try starting succulents from seed at some point, especially if you get them from your own plants. Not only will the plants be highly more likely to thrive, it's very gratifying to watch them grow under your care *(see 'Chapter 2— Make More Sucs: By Seed').*

Chapter 2
MAKE MORE SUCS

"In the autumn I gathered all my sorrows and buried them in my garden. And when April returned and spring came to wed the earth, there grew in my garden beautiful flowers unlike all other flowers. And my neighbors came to behold them, and they all said to me, "When autumn comes again, at seeding time, will you not give us of the seeds of these flowers that we may have them in our gardens?"

—Khalil Gibran, Sand and Foam, 1926

Intro

"You mean to tell me that one little leaf can grow a whole 'nother plant?"

Yep! And this is just one way to propagate your succulents. When you start a new plant by taking a cutting or a leaf, you are really making a clone, so select your hardiest, happiest specimens. A few simple preparations will increase your odds of successful propagation, but it's still a numbers game so always set out with as many leaves, cuttings, or seeds as you can handle.

Remember our little chat about growing seasons and dormancy? Good, because it's an important factor in plant propagation. If you take cuttings or start a tray of leaves in winter, you're going to be in for a long wait before you see any signs of roots. On the other hand, if you start the same tray in early spring, the results happen so quickly it's almost magical in comparison. If you've tried propagating without much success in the past, you should definitely try again during their growing season and see how your luck improves.

When propagating any part of your succulent, be sure it's hydrated. Watering a day or two before taking cuttings or leaves will give them time to uptake the water, helping ensure the propagations stay juicy long enough to put out new roots. How quickly you'll see new growth depends on the season, your climate, and the variety being propagated. Some like Graptopetalum paraguayense, may show roots in as little as a week, while others can take several weeks or longer, especially outside of their growing seasons. It really helps to have a lot of other plants to distract you while you wait. After all, it seems a watched prop never grows...

Cuttings drying in an empty pot

Propagation by Leaf

If you remove a leaf from the stem without damaging the tip, it will grow a brand spanking new plant. This clonability is seriously the coolest thing about succulents. Of course there are a few caveats: Many varieties cannot be propagated from leaves, and not all the leaves that do have this potential are going to cooperate. However, after reading this and experimenting at home, you're going to be up to your neck in propagations. Gift shopping will be reduced to needing to buy more soil and pots. Everyone will know what you're giving them for their birthday, and you sure hope they like succulents—hint hint!

All leaf propagations!
IDs clockwise from the top
fuzzy green plant:

Kalanchoe beharensis *'Fang'*
K. orgyalis *'Copper Spoons'*
Echeveria *'Perle von Nurnberg'*
Sedum adolphii *'Golden sedum'*
Echeveria *'Black Prince'*
Graptosedum *'California Sunset'*
Graptopetalum paraguayense
Sedum adolphii *'Golden sedum'*
Pachyveria oviferum *'Moonstones'*

You'll Need...

Succulent leaves: You'll need a succulent that can be propagated by leaf. You can also find online sellers who will send you a variety of leaves if you want to try your hand at different types without purchasing a bunch of individual potted plants. Either way, I suggest you begin with types known for their eagerness and ease to grow. Most genera of Echeverias and their hybrids like Sedeverias and Graptoverias, along with Graptopetalums, Graptosedums, and the larger Sedums are all a great place to begin. One commonality these share are their thicker leaves, which is often a good indicator you can propagate them. I've found there are several

I fill old nursery trays with soil for leaf propagations. They're perfect since they already have drainage holes.

colorful favorites that usually grow very well for me including Graptoveria 'Fred Ives,' Graptoveria 'Debbi,' Graptopetalum paraguayense 'Ghost Plant,' Golden sedum, Echeveria lilacina, and of course the pretty purple Echeveria 'Perle von Nurnberg.' These are just a few examples out of many.

If you're not sure whether your succulent can propagate by leaf, do a simple internet search including its name and the words 'propagation by leaf.' If you don't know its name, you can refer to 'Chapter 5—Identification', ask someone online, or just hope for the best and start a few leaves.

A tray: It should be easy to find something you already have at home to repurpose as a tray. I've used shallow cardboard boxes, large terracotta saucers, nursery trays lined with newspaper, an old drawer, a wooden wine crate, and even just the window sill. None of these items will retain water and allow decent amount of airflow around the leaves. Avoid direct contact with plastic, metal, or glass, as these can retain too much heat and prevent moisture from evaporating.

Bright indirect light: Leaves are not very hardy and the more light and heat they're exposed to, the more they will struggle to produce new growth. Experiment with where you place them a few times to get it right. I do recommend growing them outdoors but if you're growing them inside, you'll have a bit more difficulty providing them enough natural light to keep their growth compact. Either way, be careful about exposing them to full sun unless you like your succulent leaves blanched and poached. (You don't—no one does.)

Soil/Water: Do not introduce any water to your succulent leaves before they've had a chance for their tips to dry out. This includes damp soil. The leaves don't need soil or water until roots begin to form. If your climate is warm and dry, you may need

2 weeks progress—these leaves haven't had a drop of water

the assistance of some lightly dampened soil to keep your leaves plump, but stick to the rule about not getting them wet before the ends have a chance to heal. When the time comes, use your best judgment or set up one tray with soil and one without to really gauge what works best for your situation. When choosing a substrate to lay your leaves on, opt for something that dries quickly without dehydrating the leaves.

Straight pumice or Turface can be too drying, as can terracotta, so include some organic materials in your tray such as soil, coir, or Sphagnum (peat) moss.

Removing leaves

Water the plants you want to propagate at least a day ahead of time. Pluck some leaves. Put them on a tray. Wait. Wait some more. It's truly as simple as that, but removing the leaf correctly is the most important step in this manner of propagation. If the tip breaks off or is damaged in the process, it's usually going to be a dud. And there will always be duds for everyone, so don't let them get you down. Just keep at it and you'll be a nonstop propagator before you know it.

Start with a leaf from the lowest row. With one hand, brace the pot or the stem at the base so you don't pull the whole

Start with removing the lower leaves

plant out. Of course you can unpot the plant if it's too awkward to safely remove leaves without breaking them or if you already plan on topping the plant.

Lightly grip the leaf where it attaches to the stem and gently twist it left to right. The motion should be like turning a key. Wiggle it up and down, side to side, twist some more if necessary. If it doesn't readily detach, stabilize the leaf with your fingers and with your other hand, tilt the whole plant side to side, forward and back. You can also loosen one side of the leaf tip by using your nail.

Most leaf tips have a certain look when removed properly. There's usually very little sign of moisture, if any, and it retains the shape of the stem it was taken from. It's quite like when a baby tooth is ready to come out, giving little resistance and pulls away clean.

Leaves acquired! Now what?

Now all that's left to do is lay out the leaves on your tray of choice and wait. Most people place them facing the same way they grew from the stem, but they'll grow no matter what side you choose. Remember to protect them from moisture until the end of the leaf has dried. You can tell when it's safe to introduce water by examining the

Clearly a dud; never rooted

Healthy baby on fading leaf

tip that was attached to the stem. Holding it up to a light, it should look completely dry and may have even formed a callus or scab. If you plan on using dampened soil in your tray, you can start watering the soil around the leaves every few days, but don't soak it. Do your best to avoid getting the leaves wet, especially if they're going to do some time in the sun or under a grow light.

Within a week or so, you may notice some of the leaves becoming translucent. This is a sign the leaf was either damaged while being removed or was exposed to moisture too soon, but it can also happen to what seemed to be a perfectly fine leaf that hasn't even seen a cup of water, let alone been given a sip. Watch these leaves for a few more days, but be ready to toss them in the compost bin. If a leaf tip turns brown or kind of deflates and dries out, it's probably not going to be viable either. It's all good though because you already knew this could happen and have plenty of other leaves to care for. Toss the duds and move on.

While you're waiting for roots, you can have fun creating designs like mandalas with your leaves using the tray as your canvas. It's a very peaceful and pleasing way to pass the time. If you have multiple varieties propagating in the same tray, it is

Fastest rooters in the west!
Graptopetalum paraguayense, 2 weeks growth

easy to forget who is who. I've tried labeling my leaves with a soft-tipped permanent marker, but the ink fades and the leaves dry up, so take plenty of pictures to help your recall.

Growth Rate

Everyone wants to know how long it will take for their leaves to root, but there's no definite answer. The main factors are location, season, and the type of plant. Sometimes they sprout new leaves long before any sign of roots. Other times, it seems like they're rooting forever without a lick of interest in growing new leaves, and sometimes that's all they do. All we can do is watch and wait and hope.

The most impressive rate of growth I've documented was in my Graptopetalum paraguayense 'Ghost Plant.' I had taken some cuttings while cleaning up my sister-in-law's planter and wound up with over a hundred leaves. I laid them in a large terracotta saucer and kept them outside on my back porch step or in a bright south facing window depending on the precipitation that day. Within two weeks all but 12 leaves had grown new leaves, roots, or both. I'm still fully stocked with Graptopetalums even after giving away the majority of them.

I'd say the longest I've waited to see growth was after starting a batch of leaves late fall. I would have cried if I couldn't laugh at how much faster they grew the following spring. Some slower to start varieties I've worked with were Sedum rubrotinctum 'Jelly Beans,' Echeveria 'Black Prince,' Pachyphytum oviferum 'Moonstones,' and both Echeverias lilacina and 'Lola.' They tend to take longer to get grow-

This Echeveria Perle von Nurnberg propagation still has the main leaf attached.

ing, but they seem very determined to thrive.

Once stems form, if you're growing indoors or on a porch that doesn't get much wind, turn on your ceiling fan or put an electric fan nearby to mimic the natural breeze they'd be getting outdoors. Wind provides resistance that will help their stems grow thicker and stronger and trains them to grip the soil a bit tighter with their roots. Without this 'workout' their little necks can stay too thin to support their heads and nobody wants wimpy, floppy succulents.

What's Next

After the baby plants have roots, you'll be wondering what to do next. Most of the time it's best to just let them keep growing in place as long as the main leaf is still attached. It will usually dry up and fall off on its own, but sometimes they're so plump, they become part of the new plant. Once roots are established, simply treat the plant as an individual. Carefully remove it from the tray by gently loosening the soil around its roots. Pot it up, give it bright indirect light, and let its soil dry between waterings just like the rest of the gang. There are so many fun projects you can make with your tiny succulents like mini arrangements, cork planters, and temporary jewelry using Flora glue. See *'Chapter 8—Tasks & Projects'* for some ideas. Just remember that a lot of plants conform to the size of their containers, so if you want bigger succulents, give them a roomier pot.

Leaf propagation troubleshooting

The following are questions people have reached out to me with about their leaf propagations and possible solutions.

My leaves never do anything. No roots, no new leaves, nothing. After asking a few questions, it turns out they were cutting the leaves from the stem with a scalpel instead of twisting them off. This severed the leaf tip that contains the cells needed to grow a new plant. I advised them on how to remove them correctly and with a little practice, they were successful. It also turns out they were a surgeon by day...

I have a few propagations growing new leaves but no roots. This is a common occurrence and I have two theories. First, there's something going on at the cellular level that is holding the roots back. It could be damage related, biological, or both. It could also simply be a result of there not yet being a need for roots if the leaf is still full of water. I suggest refraining from moistening the soil around that leaf for at least a week or two to see if it puts out roots in search of water. Or...

All I'm getting are roots—no new leaves, just lots of roots. I do think this is more directly cell-based. If a few months pass and you're still not seeing any sign of new leaves, it might be a dud despite putting out so many roots. If you're patient, keep tending to it to see if it grows leaves—succulents are full of surprises and sometimes march to their own beat, you know?

The roots on my propagation leaves dry out and nothing else happens. Make sure the leaves aren't in direct sunlight. Some may need help burying their roots to keep them hydrated, particularly in drier climates and when growing indoors. You can prop them up with a stick or rock so the leaf tip is directed in the soil. You can also try shallowly planting the leaves up and down like carrots once the tips are dry.

My propagations are growing but their little stems are too thin to stand up. I'm in Chicago and grow my succulents indoors. This is probably due to two factors: Not enough light to keep the growth compact and not enough wind to strengthen the stems. Increase the amount of light they get and offer them a workout via a fan.

I just checked on my propagation tray and several of the leaves are turning clear or mushy. Am I overwatering? I mist them every other day. I don't like using the words 'misting' or 'spritzing' for watering succulents as this implies wetting the whole leaf, and succulents want dry leaves. It sounds like these leaves got wet while sunbathing and freaked out. When watering to moisten your soil, avoid the leaves. To water leaves that are rooted, aim a gentle stream from a spray bottle directly at the roots. Avoid allowing any water to stand on the leaves, especially if they're in the sun. Add a fan to your setup if this is a common problem.

Propagation by Division

The terms 'offsets' and 'pups' each refer to a method by which succulents reproduce new plants. 'Offsets' also describe another way succulents multiply that you'll easily recognize in Aloe Vera. Aloes and Haworthias send out underground shoots that spring up as new plants. These are also called suckers or pups, and while these terms are often used interchangeably, 'suckers' are almost always used to describe new plants that shoot up from roots or the portion of stem tucked beneath the soil. New plants that form at the base of the stem above the soil line or higher up are called 'offsets,' as with the 'hen and chicks' type of Sempervivums and Echeverias. In both cases, it's typically easier to unpot the plant to access the offsets for removal. While offsets may have already produced roots, they will still need to be treated as a cutting until the point of division has healed and callused to avoid rot.

Prepping for division

As with propagation by leaves and cuttings, working with a pre-watered plant will help keep the offsets hydrated long enough to establish roots, and the season can affect the time it takes to them grow. Sterilize any blade you may use with diluted bleach or rubbing alcohol. I prefer a straight edge for some offsets and scissors or just my fingers for others.

Unpot your plant if necessary and place it on a flat surface or in a shallow bin. Remember you'll be creating small cuts on the main plant and any water that gets in can cause rot, so plan on not watering it for a few days or more after dividing.

Top: Echeveria parva pup
Bottom: Gasteraloe *'Royal Highness'*

Hen and chicks offset division

A baby begins as a tiny rosette along the mother plant's main stem, usually at the base. It soon develops its own stem and might even start growing aerial roots that will more than likely dry out before it's time to pot the offset, but that's nothing to worry about. Look for an offset with a stem long enough to access its base without damaging its tiny leaves. With one hand, stabilize the plant just above the offset you want to propagate. Occasionally you'll be able to remove it with just your fingers by using the same technique for 'plucking' leaves for propagation. Grasp the base of the offset's stem and see if it wants to come away easily. Sometimes there's a small notch where you can slide your nail under the offset's node and detach it. Use a blade when you can't reach the base with your fingers or if it's too thick or fragile to make a clean cut without a sharp edge.

Pups and suckers division

These babies begin as new growth beneath the soil and push their way up as they mature. They remain attached to the main plant via the root system while generating their own roots. It is best to wait until they have their own roots before detaching, however many varieties can be treated as cuttings.

Place your unpotted plant in a shallow bin or on paper for easy clean-up. Inspect the offsets for health, viable roots, and maturity. If their leaves are a paler green compared to the main plant, give them some more time to mature before removal.

Follow the base of the pup along the roots to the main plant while looking for a good place to remove it. If it has its own roots, do your best to leave them intact. Some pups will pull away easily with their new roots. Otherwise, use a sterilized utensil to cut it free, and try allowing at least 1" (2.5 cm) of the main root to remain attached to it when possible.

Trim away any dead or dried roots from the main plant, and as with cuttings, allow the offsets and the main plant to completely dry and heal before potting them up, preferably in dry soil to prevent root rot. Begin watering a few days after the plant has a chance to settle in its new home.

Propagation by Cuttings

There are several good reasons why you may want to take cuttings, be it for pruning unwieldy plants, topping an etiolated rosette, or just to make more succulents, which is great because then you can share, right? Play along with me...

Many rules carry over from propagating by leaf to cuttings. As with leaf propagations, growing seasons influence how soon a cutting will take root. It's best to aim for spring through early fall, while taking a break if your summers tend to be above 90°F (32°C) and you're growing outdoors. You can still try during mid-summer, but it could be a bit more difficult to keep the plant alive long enough for it to root.

Not all succulents will be suitable for propagating by cuttings simply due to their lack of stems and branches. The following method is best suited for varieties that grow branched or trailing forms. Some examples of branching succulents include several of the varieties previously mentioned in the section 'Propagation by Leaf.' Graptopetalum paraguayense, Graptoveria *'Debbi,'* Golden sedum, Crassula ovata, and Euphorbia *'Sticks on Fire'* are good examples of succulents that form branches and often need a trim. Cuttings are perfect for managing trailing and tall-growing unruly varieties like Kalanchoe fedtschenkoi *'Lavender Scallops,'* Sedum morganianum *'Burro's Tail,'* and Senecio rowleyanus *'String of Pearls.'*

Making the cut

Prepare by watering the plants at least a day in advance. Use bleach or rubbing alcohol to sterilize your blade of choice, such as a pair of small hand pruners or scissors. You can cut at just about any point along the stem of a succulent and rightfully expect it to make a new plant. If you're working with a longer piece, you may need to turn it into a few shorter cuttings of 6" (15 cm) or less to make sure each piece has enough energy in reserve to root. Making angled cuts assists in preventing any standing water from staying on the cut stem of the main plant.

Prepping a new cutting

It helps to remove a few lower leaves to expose more of the stem so it's long enough to support the plant when potted. Do this at the time you take cuttings so everything is healing and drying simultaneously. After taking your cuttings, allow the stem to dry where it was severed. It will exhibit a callus, usually as a light brown

The dried end of a cutting looks brown, dry, and may even start putting out roots. This was cut two months before this picture was taken and never put in soil or watered.

scab, totally enclosing the wound. You shouldn't see any sign of moisture at the tip and little to no green when held under a light. Depending on the thickness of the stem, it can take three days to a couple of weeks for it to dry. I like to put my cuttings in an empty clay pot (or filled with pumice) to help them dry a bit more quickly and keep them upright. They still need good light because they continue to grow despite being recently chopped. Applying powdered sulfur to the cut can help prevent infection while it heals. Lower leaves remaining on the main plant will lose their shade and may be sensitive to the new amount of light for a while, so give them a bit less light until they adjust. In time, new growth should form on the main stem where it was cut and make you a whole new plant!

Potting a cutting

Once your cuttings are dry, you can pot them in dry or lightly dampened soil to help support them and wait for them to take root. Ideally, wait until you see roots before 'really' watering. This can take anywhere from a week to a few months, so water them with a lighter hand than you would rooted plants. In my experience, cuttings tend to root more quickly when deprived of plentiful moisture, as if they know they're going to need to grow roots soon if they want a drink.

Troubleshooting cuttings

The only thing you really need to look out for is watering a cutting before the end is fully healed. If you start to see the cut end turn soft, translucent, or brown, there's a chance the stem wasn't thoroughly dried before it was watered. The good news is you can still try to save it by cutting above the rotted portion until you reach healthy stem again. Then repeat the prepping process and try again.

If a month passes and your cuttings haven't taken root, you can try a few things. Applying rooting hormone to the dried end and repotting might stimulate growth. You can also try water propagation by filling a glass bottle with enough water to reach just above tip of the fully callused stem. Finally, you can always cut a bit higher up on the stem, let it dry, and try the old fashioned way again.

Crassula ovata easily
propagates by cuttings

Cactus Cuttings

To propagate columnar cacti by cuttings, simply make a clean cut and keep the plant healthy long enough to produce new growth. It's much easier to cut a cactus while it's unpotted. Lay it on its side, stabilize it with one hand, and make your cut with a sharp sterilized blade. Allow the cutting to dry before potting in dry soil and wait for several weeks to water. Repot the main plant, but don't water it for at least a week. The new plants will form along or near the cut.

For leafy epiphytic cacti, such as Epiphyllum oxypetalum 'Queen of the Night,' cut the leaf as close to the base as you can and let the wound dry thoroughly before tucking it into a pot of dry soil. In time it will form roots and may even begin to put out new growth from the edges of the leaf.

If you have a cactus that produces offsets, they often send out their own roots while still attached to the main plant. You can try to gently twist the babies off or use a small blade to free them while doing your best not to disturb their young roots. Even if they don't have their own roots, you can still remove them but it's very important not to water the offsets until they have several roots established.

Cereus peruvianus new growth; above photo was taken in May, right photo was taken October.

Propagation by Seed

Caring for seedlings requires more frequent watering, good light, warmth and a lot of patience—all well worth that first glimpse of green poking out of the soil, I promise. And it's easier than you may have imagined if you have the right setup.

If you don't have a grow light and you want to try starting succulent seeds, now is the time to consider buying one. They guarantee warm temperatures and good light year round, and you'll need those elements to keep your seedlings strong.

Besides needing consistent good light, seedlings want warmth to germinate and flourish. About 70°F (21°C) is ideal. Finally, unlike the mature plants, succulent seedlings like humidity and need to be watered frequently enough to keep the soil from drying out while they are becoming established plants.

To ensure timely germination, start your seeds a couple of weeks before spring. I started some cacti seedlings late winter and waited two months before seeing any growth, but when I started another batch in spring, they germinated in two weeks.

Harvesting seeds

Some succulents produce more obvious seed pods or fruits that will be easier to harvest than others. Most will require cross-pollination but many are self-fertile. Allow the flowers to dry on the plant to give the seeds time to mature. This amount of time will vary based on the type of plant. Some seeds can be stored for a year or more while others should be planted while still fresh.

Varieties like Echeveria form tiny pods at the base of the dried flower and the seeds are super small. You can gently crush the dried pod and sprinkle the seeds on top of soil or store the pod in a cool, dry location.

Cacti, Lithops, Stapelia, Aloes, and Haworthias are some examples that form larger pods or fruits that are much easier to harvest. Cacti fruits should be harvested while they are still fresh or the pod will dry up and make the seeds harder to extract.

The photo below shows a Mammillaria fruit after being cut open. You can wipe the split fruit on a piece of paper and once the pulp is dry the seeds easily scrape off. Fold the paper and use the crease to help guide the seeds into a storage bag.

Macro photo of a Mammillaria elegans seed pod. The fruit will easily pull away when ready to harvest.

Mammillaria schiedeana fruit

Aloe maculata seed pods

Lithops seedlings, 10 months

Faucaria seedlings, 6 months

Instructions on Starting Succulent Seeds

Supplies:

Seeds
Soil– Seedling mix or your basic succulent soil
Containers– No glass or metal, must have drainage holes
Sifter– For the top layer of soil
Tray– As a saucer for catching water
Fine water mister– To water without disturbing soil

1 Repurpose your to-go containers. I put the container on a few towels and poked the holes with a carving fork. Poke holes from the inside out so they'll drain outwards.

2 Smooth the soft sand layer over the surface.

3 Use a spritzer bottle to thoroughly moisten the soil.

4 You can mark the soil with shallow holes to keep things organized.

5 Drop a seed in each hole. If they're really small, place them on top of the soil.

6 Gently tap next to each hole to cover the seeds then spritz the soil again.

Choose and prepare your container

Until your seeds germinate, you'll want to cover the container with plastic to create humidity, so keep that in mind when deciding what to sow in. The covering shouldn't be airtight so poke some holes in plastic wrap or leave most of a zipper bag unsealed. You can use a seed starter tray or repurpose a number of other containers you may already have on hand. I drilled some holes in an ice cube tray which was small enough to slip into a gallon plastic zipper bag. Clear lidded to-go containers with added drainage holes are also useful and abundant. You can also use small nursery pots or flats filled with soil.

Fill the container with soil allowing room for another 1/8" (3 mm) layer of soil to cover the seeds. Water until the soil is thoroughly moist. Use the sifter to make enough 'soft sand' to cover the seeds. This removes large particles that might hinder

Euphorbia polygona 'Snowflake'
seedling progress: 4 months to 2 years

the growth of your seedlings. Leave another 1/4" (6 mm) of space between the top rim and the soil to keep water from spilling over.

Planting seeds

Try not to overcrowd the container with too many seeds. Spacing them an inch or two apart (2–5 cm) will make it easier to transplant them down the road. Place them on top of the moistened soil and cover them with the sifted soil. Then using your spritzer, carefully dampen the top layer being careful not to uncover the seeds. If they are super tiny, like Lithops seeds, don't cover them with soil or spray them or they'll scatter. You can also soak the pot from the bottom until the top layer is saturated. Cover the container with your choice of plastic and place on a tray to catch any excess water. Continue to use the spritzer to gently water your seeds or bottom-water for the next several waterings until the soil has time to settle. Water when the top layer of soil begins to look dry.

Damping off

Damping off is a disease that occurs when seeds or seedlings are attacked by pathogens in the soil, causing them to rot and die. Many growers take the extra step to sterilize the soil with heat until the 'bad' fungi and bacteria are killed off. Others use cinnamon, chamomile tea, diluted hydrogen peroxide, or other home remedies to kill or discourage fungal growth. To sterilize your soil you'll need to heat it to 180°F (82°C) for half an hour. Whether you use your oven, stove top, microwave, or pressure cooker, be sure to moisten the soil so it doesn't burn.

POP!

Before long, typically between 10 days to 6 weeks, you will see the first sprouts emerge. They won't all germinate at once and it's common for some seeds to fail, but it is a very exciting time when you first notice new life popping up through the soil. I hope you take some progress shots of their growth because it's going to be a good while before you can do anything else with them. Most varieties take a couple of years to reach transplant size, so be prepared to wait and to lose a few weaklings. Once they're of size to be in their own pot, carefully uproot them with a spoon and replant them in a larger container of your choice, but skip watering for up to two weeks to allow their roots to heal. Let them settle in for a few months at least before moving them again to avoid transplant shock.

While growing from seed takes much longer than other forms of propagation, it is a good experience for you to really get to know your plants. The next challenge would be to try cross-pollinating. For that you need two types of succulent flowers in bloom that will cross and a good deal of patience. While it is as simple as moving pollen from one flower to another as a bee would do naturally, it can take over a year to see if the results produce viable seeds or any seeds at all. If you have the time and setup to perform this experiment and you succeed, you will gain a whole new level of wisdom and respect for your plants. Take a lot of pictures and notes to keep track of the process so you can replicate the steps and share your knowledge with others.

Chapter 3
SUCCULENT SOS

"My green thumb came only as a result of the mistakes I made while learning to see things from the plants' point of view."

—H. Fred Dale, Toronto Star garden writer and author of Fred Dale's Garden Book, 1972

Mealybugs hiding between an Echeveria's leaves

Succulent SOS: Symptoms

The following **Symptoms List** details some of the most common signs of problems found in succulents and their possible reasons. The next section in this chapter, **'Take Action,'** provides instructions on how to correct the issues.

While some symptoms are similar they can indicate different causes, like how wrinkling in leaves can be due to overwatering or a need for more frequent watering. In cases like this, you'll need to assess the amount of light and water your plant has been receiving, how fast the soil is drying, and the climate, including temperature and humidity.

The most frequent problems are solved by watering less, offering more light, or a combination of both. First make sure you're meeting the basic needs for your succulents. You can refer back to the **'Basic Tips'** sections from Chapter 1 below for more information.

Soil: You're using a high-draining, low organic soil
See 'Soil and Fertilizer'

Light: Your plant gets enough light to keep it growing true to form
See 'Light'

Water: The soil is drying out within a week of the last watering
See 'Watering'

Pot: The pot has good sized drainage holes and its size fits your plant
See 'Containers'

If you said yes to all of the above, you're ready to begin troubleshooting. Refer to the suggested solutions in the next section **'Take Action.'**

Symptoms List

Signs in Leaves
Signs in Stems
Signs in Roots
Signs of Fungal Diseases
Signs of Common Pests

Signs in Leaves by Texture, Behavior, and Coloring

The first sign of a troubled succulent tends to show in its leaves, so we'll start there.

By Texture

Soft leaves, as in not as firm when gently squeezed: Soft leaves may be a sign of too much moisture being held in the soil for too long or a need for a drink. Check drainage and drying rate of soil before gradually increasing watering frequency. Heat can also be a factor.
See 'Increase Hydration' and 'Reduce Hydration'

Wrinkled leaves: Leaves can wrinkle as they use up stored water or after getting too much sun. This usually occurs more in hotter seasons. Gradually increase watering frequency and consider if less light may help. However, if wrinkled leaves also look droopy, translucent, or fall off, overwatering is more likely the cause.
See 'Increase Hydration' and 'Reduce Hydration'

Anacampseros rufescens; note how the leaves look wrinkled. They also feel less firm than when fully hydrated.

By Behavior

Droopy, sagging leaves: Leaves can get too heavy when overwatered. They may also display more of the surface area when they want more light. If most of the surrounding leaves look happy, inspect closely for signs of pests or fungi.
See 'Reduce Hydration,' 'Increase Light,' 'Pests' and 'Fungi'

Leaves dropping: While older leaves do drop naturally, new leaves dropping is another sign of overwatering which weakens the spot where leaves attach to the stem. It can also mean your plants are under attack by something like mealybugs or fungi, or that stem rot is present.
See 'Reduce Hydration,' 'Stem/Root Rot,' 'Pests,' and 'Fungi'

Leaves curling in or thinning: Aloe and Haworthia are two examples of plants that really show what they're feeling through their leaves. Aloe leaves flatten out and get soft when thirsty, while Haworthia leaves usually stay firm but begin to look concave. Aloe and Haworthia will both curl along their leaf margins when in need of hydration.
See 'Increase Hydration'

New leaves growing in are spaced further apart on the stem; plant heads turning towards the light; plants stretching towards the light; leaves curling under or spreading open wider: An increase in spacing between the leaves of succulent varieties that form rosettes like Echeverias or stacked leaves like Crassula perforata 'String of Buttons' is a tell-tale sign of etiolation caused by a need for more light. Plants may lean and grow towards the light source more than normal, or their leaves may curl under or spread wider to maximize exposure to light.
See 'Etiolation Repair' and 'Increase Light'

New leaf growth looks stunted and/or deformed: Pests like mealybugs target the tenderest leaves, which is often the newest growth. The new leaf may look unsymmetrical or seem to be struggling to grow.
See 'Pests'

Older lower leaves drying out can be normal behavior

Old growth vs new: E. Lola after correcting the light

Graptopetalum superbum leaves spaced too far apart

Translucent Sedum *'Jelly Beans'*; these died after a light freeze

Greenovia displaying lighter leaves on new growth

Scorch marks from water on leaves left in direct sunlight

By Coloring

Leaves turning translucent: If it hasn't been too cold recently, this is most certainly a sign of overwatering and possible rot.
See 'Reduce Hydration,' 'Stem/Root Rot'

Leaves turning yellow: This could be a symptom of too much direct light, heat, or a need for water. It can also be due to being too cold.
See 'Increase Hydration,' and 'Decrease Light'

New leaf growth is lighter than old: In some cases this is normal but it is also an indicator of a need for more light, especially when accompanied by the older leaves spreading and flattening out as many Echeverias or Sempervivums are quick to do. Gradually increase light to improve shape and color.
See 'Increase Light'

Colorful leaves are turning green: As mentioned in the first chapter under *'Basic Tips: Color and Form,'* succulents often lose their brilliant hues not long after you bring them home because they are being relocated to settings that don't have the ideal light, temperature, and climate a controlled greenhouse provides. It is normal, but some varieties can be coaxed into keeping their colors.
See 'Color Repair'

Spotting or Scars on Leaves

'Spots' is a vague term. Here I am referring to fixed damage to the plant and not from pests like mealybugs or mineral deposits that can be wiped off. If you're unsure, moisten a cotton swab and try to remove the spot. Spots can signify many things like overwatering, sunburn, cold damage, fungal or pest issues. They can be raised, flat, or concave. While I've added some appearance details, the color, size, and shape can vary.

Small brown spots scattered across the leaves like freckles are a sign of edema (odema), where a cell has burst from too much water. They can be flat, raised, or concave.
See 'Reduce Hydration'

Large brown spots on top of the leaves or nearer to the stem, are often a sign of a drop of water sitting on the leaf in bright light and heat for too long, or a larger section of edema damage. Usually flat or concave, they may turn white before scarring. Take action by waiting until closer to sunset or after dark to water your plants and doing what you can to dry them off if caught in the rain on a sunny day. You can shake off the droplets, use a turkey baster or just your breath if there aren't many plants to worry about.

Aeonium *'Kiwi'* showing scars from its own leaves

Concave spots and scars can mean something was nibbling on your plant or it was physically damaged in some other way. For example, Aeonium 'Kiwi' are very sensitive to damage and often show brown scars from their own leaves or neighboring plants.
See 'Pests' and 'Physical Damage'

Raised spots and scars can be caused by mites, scale, and edema from overwatering.
See 'Pests' and 'Reduce Hydration'

Signs in Stems

Darkened Echeveria *'Chroma'* stem with rot

Stems becoming dark, translucent, or soft, typically close to the soil and below the soil line are all symptoms of stem rot or root rot. This tends to happens when water or bacteria enters the plant through a wound in the roots or stems. Wounds can be caused by pests, while repotting, or when removing leaves to propagate or taking cuttings. Overwatering and poor drainage can also suffocate roots which die and decay, creating a bacterial breeding ground.
See 'Stem/Root Rot'

Circled: Mealybug found during a cutting root check

Black sooty mold is a fungus which resembles dirt or soot

Macro shot of mealybugs hiding on a leaf

Woolly aphid on a cactus spine

Signs in Roots

Roots drying out and falling off can be a sign of a need for more water, overwatering, poor soil, root mealybugs, or something else related to the health of the plant. Sometimes this is normal and part of a plants' growing cycle, particularly in those which go through a dormant period when the roots are temporarily put out of service. However, if your succulent is obviously unhealthy and you're unsure of condition of the roots, you'll want to unpot it and try to determine the cause.

See 'Stem/Root Rot,' 'Reduce Hydration,' and 'Pests'

Signs of Fungal Disease and Common Pests

Fungi

Black Sooty Mold: If you see something on your leaves that looks like dirt but it doesn't easily rinse away, it could be black sooty mold, a fungus that feeds on the honeydew secretions of insects like aphids and mealybugs, as well as flower nectar. Spores are spread by insects, water drops, and wind.

See 'Fungi'

Powdery Mildew: This white powdery fungus looks just like its name and feeds on the epidural cells of a plant. The spores are typically transmitted by wind or water.

See 'Fungi'

Pests

Mealybugs: While there are nearly 300 types of mealybugs in the U.S. alone, they are typically very small, flat, oval-shaped and white to light yellow in color. They hide down in the leaves near or on the stems and on the underside of leaves. They can be seen individually or grouped with others. Only the adult males are able to fly, but they're very small and short lived so you'll probably only see the females. They're often seen near a white flaky or cottony egg mass along the stem or near the leaf nodes. Sometimes they look like a small piece of perlite or pumice. You can verify their presence by poking the bug with a Q-Tip, pencil tip, or soft brush and looking for signs of blood ranging from yellow to orange. (Sorry, I know that's pretty gross.)

Some mealybugs live among the roots and will only be noticed if the plant is unpotted. They are usually smaller than the leaf-dwellers and leave behind a cottony or powdery white substance. Be careful to inspect any new plants you bring home for pests before putting them near your other plants, as mealybugs can crawl or get carried by ants to the rest of your collection.
See 'Pests – Mealybugs'

Aphids are more active and usually larger than mealybugs

Aphids: Aphids are easy to confuse with mealybugs, but they are usually a bit larger, darker, and more active than mealies. They're also more likely to visibly congregate along a stalk or on a flower rather than hide tucked down in the leaves.
See 'Pests – Aphids'

Spider Mites: Spider mites are incredibly small and range from black to red if you can even see them. Their webbing is much easier to spot and you'll probably notice it first. Spider mites weaken plants by feeding on their juices. The damage usually looks like small yellow to brown spots.
See 'Pests – Spider Mites'

Diaspis scale on Opuntia pad

Scale: More common to cacti than soft succulents, there are two main types of scale insects to watch out for. Diaspis echinocacti scale females create small, dark to light brown circular spots with a distinctively raised nucleus. They look like flattened barnacles. The males are extremely small, elongated white crawlers.

Cochineal scale (Dactylopius coccus) is soft-bodied and targets Opuntia. Females are small, flat, and oval in shape, and often exhibit a waxy white or cottony coating. Males are smaller, winged, and rarer to see.
See 'Pests – Scale'

Cochineal scale on Opuntia pad

Fungus Gnats: Gnats are typically more noticeable in indoor plants. The small flying adults are harmless, but they lay eggs in moist soil where their larva feeds on fungus, root hairs, and other organic matter. Eggs and larvae cannot survive in dry soil.
See 'Reduce Hydration'

Ant 'farming' a mealybug on a Golden Sedum

Caught in the act: A caterpillar munching on a Kalanchoe leaf

Holes and frass: Obvious signs a caterpillar was here

Black Carpenter Ants and others: While these tiny ants don't harm your succulents directly, if you see them on your plants, you need to take a closer look and inspect for mealybugs. These are a type of sugar ants that farm mealybugs and aphids for their sweet secretions, aptly named 'honeydew.' These ants will not only protect mealybugs from other insects, they are often the very reason for their presence.

Sugar ants deposit mealybugs and aphids on plants for the purpose of coming back to harvest their secretions, but they don't kill them. Instead, they agitate them—it looks like they're tickling them—which causes the mealybugs to secrete their juices. The ants will also relocate mealybugs to different plants. The honeydew makes an attractive food for black sooty mold.

So while these ants are part of the problem, they're usually what leads me straight to the mealybugs. They are also pollinators, so I'm a bit turned off from killing them. However, your best bet for getting rid of these ants is to find their nest and destroy it.

Slugs and Snails: These night feeders may be slow but they can devour succulent leaves with great speed. If you see small holes and chunks missing from your plants' leaves, it's likely from a hungry snail, slug, or caterpillar. Traces of shiny slime trails and excrement that looks like fish poo—longish, skinny brown droppings—are sure signs of snails or slugs feasting on your plants.
See 'Pests – Slugs and Snails'

Caterpillars: Most of the caterpillars I've found on my succulents are from moths that got trapped on my porch and laid their eggs on the plant. Signs of a caterpillar include chunks missing from leaves, leaves missing from branches, thin webbing, and the dead giveaway, caterpillar frass (poop) which looks like dark round balls about the size of a poppy seed.
See 'Pests – Caterpillars'

Other garden pests: Rodents, wasps, rabbits, raccoons, and birds may find your succulents tasty. Look for uprooted plants, chunks of leaves missing (or entire leaves), and holes and scrapes on the surface of leaves.
See 'Pests – Birds and Mammals'

Take Action

Increase Hydration
Reduce Hydration
Increase Light
Decrease Light
Color Repair
Etiolation Repair
Stem Rot
Root Rot
Fungi
Pests
Mealybugs
Aphids
Spider Mites
Scale
Slugs and Snails
Caterpillars
Birds and Mammals
Physical Damage

Rescuing an offset from an Echeveria 'Chroma' with stem rot from too much water

Increase Hydration

If you are watering less than once a week and your soil is drying out between waterings, your succulent could need more frequent watering. You should see some water draining through the pot holes within a minute of watering but not too much. Start watering every 4–5 days instead of weekly. Be careful to avoid overwatering during winter since many succulents go dormant. More water can stimulate growth while sunlight is insufficient, which can lead to etiolation. The soil also takes longer to dry, which can lead to rot.

Reduce Hydration

Under most conditions, weekly watering gives the soil time to dry out and maybe stay dry for a day or two. If your soil isn't drying out within that time or your pot is too large for the plant, excess moisture can invite bacteria and fungi, as well as fungus gnats.

First, make sure your soil is draining quickly and the pot isn't hindering evaporation (see 'Chapter 1—Basic Tips: Containers'). Reduce the amount and frequency of water to once every 10 days or try watering less deeply but more frequently.

Increase Light

When more light is prescribed, you'll need to acclimate your plant to an increase in the hours and brightness of the light or you risk scorching or blanching the leaves. If you are keeping your plant potted, move it to a location where it will get more light for a couple of hours a day or less for about a week. Remember that direct morning sun is gentler than afternoon sun. Increase the amount of time they spend in brighter light over the course of a week or two while paying attention to their reaction and inspecting for any signs of sunburn. If you see any leaves lightening, they could be getting blanched from too much sun too soon. Darkening at the tips can indicate burn or excess water in the leaves.

If you are going to put your succulent in the ground, use the same method as a potted plant to acclimate it to its new location until it seems happy, then go ahead and plant it. Keep an eye on it though and be ready to dig it up if it starts to struggle. If you don't have sufficient natural light, there are always grow lights *(see 'Chapter 1— Basic Tips: Light').*

Decrease Light

Most succulents thrive when light is bright but indirect. String of Pearls and Burro's Tail are common examples of varieties that prefer part sun over full sun. Reduce watering while decreasing light to slow the growth of the plant and avoid etiolation. The soil will take longer to dry, so add more drainage materials or holes in the pot, and water less if need be.

Color Repair

While it's normal for succulents to lose their brilliant colors once they're no longer kept in a greenhouse *(see 'Chapter 1—Basic Tips: Color and Form')* there

The lighter new growth in this Echeveria shows it wants more light

This etiolated Echeveria should be given more light before being topped

Edema scars due to overwatering or direct sun on wet leaves

are a few ways you can encourage them to return. When temperatures warm up, we naturally want to shade or water our plants more often, and many succulents need this to make it through the heat of summer. This reduces the stress and causes succulents to revert to their unstressed colors. Bright light, cool temperatures, and a reduction in water are usually all necessary for stressing a succulent to produce its colors in non-arid climates.

I watched my outdoor Echeveria purpusorum 'Dionysos' lose its purply markings when summer settled in so I moved it to a bright east-facing window sill indoors where it's much cooler. The purples returned in only a few weeks.

Grow lights are really the only way to guarantee your succulents get a daily dose of the light required to maintain these colors, but nature will kindly provide the right conditions depending on where you live. The cool yet bright days of early spring really bring out the vibrant hues, and most succulent keepers agree it's their favorite time of the year to watch the colors return.

An unstressed Echeveria pulidonis vs. a beautifully stressed specimen

Etiolation Repair

Increasing light will help new growth return to its ideal form, but eventually you may want to 'top' the plant by cutting off the top portion to reduce the length of the stem. The method for topping a succulent is the same as taking cuttings *(see 'Chapter 2—Make More Sucs: By Cuttings')*. While it isn't always necessary and usually only an aesthetic choice, etiolated stems can become too weak to support the top growth of the plant and still remain upright. However, certain varieties will spill and trail over the edges of pots. They will look quite lovely without being topped as long as they're receiving adequate light.

Cut above the rot until the stem looks healthy again

Allow the cut to callus over before introducing water or damp soil

Cactus with stem rot drying after being treated

Stem Rot

It's possible to save a succulent from rot by taking cuttings above the unhealthy portion of the stem. There should be no sign of brown coloring or translucent appearance at the point you cut. It will look green, firm and healthy. This is easier to manage in succulents that have visible stems, but it's possible to repair plants like Aloes, Haworthias, and even cacti in this manner.

In Aloes and Haworthias, you'll need to unpot the plant and start removing the lowest leaves until you reveal the stem. From there you can determine how high up you need to cut to remove the unhealthy portion. As these stems are much thicker, they will take longer to dry. Be sure to use sterilized clippers or else the bacteria will spread to the healthy cutting.

In rescuing cacti from stem rot, use a sharp sterilized knife to make a clean cut above the rotted portion. Let the cut end dry for several weeks before potting in dry soil, then don't water until you see roots. Be sure to keep it out of direct sun. This isn't always an option in thicker cacti but for the taller, columnar varieties, there's still a fighting chance, especially during their growing seasons when they're more likely to root.

Root Rot

Unpot your plant and brush away as much soil from the roots as you can. Give them a gentle tug to feel for resistance. If they come away easily or seem dry, moldy, or rotted, you can try removing the roots with sterile clippers or go ahead and treat the plant as you would for stem rot. A lot of times, you'll notice the root system is almost non-existent or very shallow and sparse when it should be deeper and fuller.

To try rerooting your succulent, rinse the roots of old soil and let the plant dry. Cut away all the dead or dying roots until you can see the base of the plant. Let it dry in an empty clay pot stuffed with paper while monitoring the health of the leaves and any stem you can see. If it looks okay, you can set it on a pot of loose dry soil where it will begin to put out new roots. Do not water until new roots form or else the base of the plant can rot. This can take a month or more. If the plant doesn't root, the rot may have spread up the stem. Proceed to treating it for stem rot.

Fungi

Flower nectar and the honeydew secretion of mealybugs and aphids are feeding and breeding grounds for black sooty mold. High humidity and tight quarters facilitates powdery mildew growth.

Treating Black Sooty Mold: First be sure you've washed away any traces of nectar or honeydew with diluted water and rubbing alcohol or a high-powered water hose nozzle if your plant is a more durable variety. Prune and discard any leaves that don't come clean. Stay vigilant about keeping your plants free of any honeydew secretors like mealybugs, aphids, whiteflies, leafhoppers, and scale. While sooty molds don't directly infect succulents, they can weaken their defenses against other pests and hinder photosynthesis.

Treating Powdery Mildew: Make sure you haven't misidentified farina, the natural powdery coating on succulents, as powdery mildew. Try spraying the mildew from the leaves with as strong of a stream of water your plant will tolerate. If this isn't effective, you can try spraying them with a solution of baking soda and water. Mix 3 tablespoons of baking soda in a gallon of water then add a few drops of liquid soap to help it adhere to the leaves. Spray thoroughly but do so at a time when the sun won't be directly on the plants or they could get scorched. Rinse well after a few hours and repeat as needed.

Prevent powdery mildew by potting in a way that allows good airflow around your plants. It thrives in shady areas and when temperatures are mild. Since succulents are kept in bright locations, it's more likely to infect other plants first. Most powdery mildew spores stop germinating at temperatures above 90°F (32°C).

Pests

Mealybugs

Due to their weeness, these pests can be real punks to see and squash. You'll have better odds of finding them all if you unpot the plant you're treating, and you won't risk them falling off into the pot only to crawl back up and hide again. You should also inspect the roots for mealybugs or other problems while it's unpotted.

First, quarantine the plant apart from your other plants until you can tend to it and while treating it so the little busters don't escape and relocate to another plant.

Fill a squirt bottle with 3 parts rubbing alcohol and 1 part water. Hold your plant over a container while spot treating the mealybugs until you are satisfied that you got them all. More than likely, there are still babies or eggs that you missed so you can fill another container with the same solution and submerge your plant upside down in it. Gently swish and swirl the plant around to make sure the solution gets down into all of its nooks and crannies. Rinse with fresh water and repeat until you no longer see any mealybugs.

After treating your plants, keep them out of direct sunlight. You can leave the plants submerged in the solution with the roots out of the water for about 30 minutes to make sure the mealybugs are dead. They are incredibly tenacious and can survive underwater so do your best to make sure they're all gone before repotting. I usually wait a day or two and inspect them several times before putting them back in pots. It's also probably a good idea to replace the soil in case any pests are still lurking about.

If you're still seeing mealybugs after treating them, you can try using full strength alcohol, other topical solutions like Neem oil, or go for the big guns and use systemic pesticides that work from the inside out. Ladybugs are sometimes helpful in mealybug control, but due to how small mealies can be, they're often overlooked by this beneficial beetle.

Aphids

While you can treat your plant for aphids in the same manner as mealybugs, they're a bit easier to detect and much easier to kill though you may have to repeat the treatment to thoroughly remove them all.

Top: Swirling succulents in water after an alcohol treatment to dislodge pests

Bottom: Using alcohol and a paintbrush to remove scale around an areole

Ladybugs and lacewing larva are incredibly effective at consuming vast numbers of aphids. Lacewing larvae are the weird little insects that build mobile homes of decayed organic material on their backs. With their fierce looking 'jaws,' they look a lot like ladybug larvae. You've probably seen them before but may not have realized how they're helping keep your plants pest-free.

Adult lacewings are bright green with delicate clear wings. Their eggs are very distinctive—tiny white ovals suspended on a delicate web-like stalk, and often laid in a spiral formation.

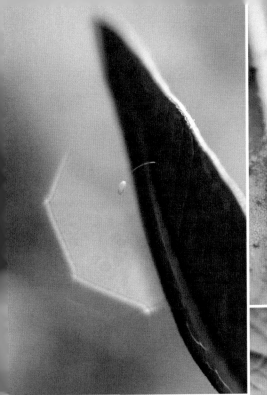

Know your garden friends!

Top left: Lacewing egg
Bottom left: Mature lacewing fly
Top right: Ladybug eggs
Middle right: Ladybug larva
Bottom right: Ladybug with an aphid

Spider Mites

The simplest way to avoid and remove spider mites is to water your plants from overhead. While this isn't ideal or possible with some succulents, it is fine for most as long as they're not in full sun while drying. Systemic insecticides don't work on spider mites since they're arachnids, not insects, and can actually encourage their reproduction by killing off their natural predators and stimulating leaves to produce more nitrogen, the main nutrient leaf-sucking pests thrive on. Topical treatment with insecticidal oils or soaps are effective but must make contact with the mites to kill them and needs to be rinsed thoroughly from leaves to prevent further damage.

Scale

Scale can be hard to remove with water alone. Applying alcohol prior to spraying or scrubbing the scale off helps loosen it up. Use a toothbrush, cotton swab, or a strong stream of water to blast away the tiny flat pests.

Rinse the plant thoroughly and be prepared to repeat the treatment the following day. You can leave the plant unpotted for a few days if you plan on treating it again. In particularly bad cases of scale, you may need to use a systemic soil drench which works from the inside out, however systemics are toxic and not pollinator-friendly so cut off any flowers if you're keeping them outdoors.

Slugs and Snails

Snails and slugs are usually nocturnal feeders and I have to say it's quite exciting to get out my flashlight and go hunting for them at night. They also come out after a good rain during the day. I don't kill them because they're just doing their snail thing, and beneficial toads along with birds like robins and starlings love them. My friend says her chickens love them too. I put my bounty in a jar and relocate them away from my house or just flick them as far as I can. You can trap them in a lidded jar and put them in the freezer if you want to kill them humanely. Dropping them in a bucket of soapy water is less humane but gets the job done.

There are a few methods you can use to trap or deter snails and slugs such as applying copper tape strips along pots and garden beds which gives them a little zap when they make contact with it. Rough sandpaper placed under your pots will be too coarse for them to travel across and may help. I've made a trap from an inverted 2-liter bottle filled with lunch meat. It was disgusting, but it worked. Some people use saucers of beer to lure them as well.

Caterpillars

Moths, butterflies, and other larva-producing insects may lay eggs on your succulents. If you see any of these on your porch or in your home, catch and release them away from your plants. Caterpillars can be tricky to spot thanks to their camouflaging skills. Look for them along the leaf edges, under leaves, and along stems. If you find one, take it outside away from your plants and set it loose. If it doesn't get eaten by a predator, it will be far enough away to not need to worry about it returning.

This is why you shouldn't store peanuts on your screened porch... squirrels are very fast learners!

Birds and Mammals

I get a lot of visitors in my backyard—birds, squirrels, mice and rats mainly—and on a rare occasion, I see signs that one of them has been nibbling on my succulents. Often they're just looking for a drink. It seems they only take a couple of bites before deciding it's not their cup of tea, but I have heard many reports of them causing much more damage to others' plants.

I set out a water dish and more delicious treats like peanuts and seeds, and while people will argue that only attracts more critters, I haven't see an uptick in traffic.

Plastic snakes and owls, streamers on sticks, scarecrows, pepper spray, dried blood, motion activated devices...the methods of scaring away wildlife are endless, but few are effective for long.

Animals are smart and once they figure out the threat isn't real, they'll resume eating your plants. Netting and other physical barriers are really the only sure method of keeping birds and mammals out of your pots and gardens.

Physical Damage

If your succulents are damaged, they're at risk for being infected by bacteria that causes rot. Remove broken leaves promptly and trim back any stems that were partially severed. You can try propagating the pieces you take off. Larger wounds can be treated with a coating of sulfur powder to help prevent infection.

Leaves that have suffered severely from sun damage will not heal and should be removed so the energy being put into keeping them alive can be redirected to new, healthy growth.

Avoid potting tender-leaved succulents next to spiky plants like Aloes, Euphorbias, and cacti since they can be difficult neighbors.

Chapter 4
REGIONAL TIPS

"It was one of those March days when the sun shines hot and the wind blows cold: when it is summer in the light, and winter in the shade."

—Charles Dickens, Great Expectations, 1860

Intro

As the number of succulent lovers I connected with grew, I began to get a lot of questions from others about solving problems they were having with their plants based on their climate. Knowing your garden hardiness zone is important, however both California and Florida have 9A zones, yet the East Coast Sunshine State's annual rainfall is double that of the entire Golden State. Some states and countries embody multiple climates and microclimates across their geographies, and many regions see two or more widely differing climates throughout the year. So rather than referring to zones, I will use more specific climates and scenarios to explain the perks and problems you may encounter in your region.

Lack of Sunlight: Influenced by climate and latitude

Hot—Humid: Subtropical to tropical, hot and humid summers

Hot—Arid to Semi-arid: Desert to Mediterranean, hot and dry summers

Cold: Frost, freezing, snow, temperatures below 60°F (15°C)

Mixed: Combination of the above

Indoors: Temperature controlled, sheltered from precipitation

I wanted a visual of where people are keeping succulents around the world so I put out a call for anyone interested to reply with their location which I could then add to a public map online. I also hoped it would help them find nearby like-minded members of the succulent community so they could reach out to each other with questions and advice based on their specific regional challenges.

The entries poured in and it became very clear that succulents are an international passion. From Canada to Philippines and everywhere in between—except for Antarctica...I'm still waiting to hear from someone there—succulents have found a way into the homes and hearts of millions, and people are not letting their difficult climates get in the way of their love for these plants.

While some are destined to keep their collections indoors or in greenhouses year-round due to the cold, others face extreme heat, humidity, rainy seasons, and other challenges that may also require an indoor setup throughout the year. If you can offer a bright, temperate shelter to your succulents when the weather forces you to bring them inside, even if only for the winter or the peak of summer, nothing is stopping you from growing them. Well, you might need more shelf space but that's an easy fix.

Lack of Sunlight

Examples: Overcast regions, high latitudes, growing indoors without grow lights

Challenges: Keeping succulents true to form, lack of flowers, soil not drying fast enough

Suggested genera: Dark green Haworthias and Aloes; these will still need 5+ hours of good light

Difficult genera: Cacti, Echeverias, Mesembs, lighter colored and variegated succulents

I polled my social media followers on what their biggest regional challenges to growing succulents were and the majority had issues with needing more sunlight. Those from arid states, such as Arizona and California, had the opposite challenge of too much sun. If you're in the first group, do any of these sound familiar?

–It's too hot or too cold outside for your succulents so you have to keep them indoors where the light inside isn't that great.

–You live in a location that seems to have more overcast days than sunny.

–You don't have an outdoor growing area protected from rain that gets good natural light.

I'm guessing that one or more of these scenarios describes your situation, as all three of these fit me at some point during the year. There are several solutions but none without a bit of labor and a price tag. I would never advocate against you trying to overcome any roadblock to growing succulents because I think it's all part of the lesson we learn along the way. In fact, that's exactly what Sucs for You was born of: a love for succulents too great to accept defeat—heat, rain, and humidity be damned.

If you love them too and want to grow them despite your difficult climate or location, you are accepting the role of Mother Nature. You'll need create a home for succulents in an environment that may be exactly the opposite of where they are native to. So how can we turn a small Brooklyn

Haworthias and 'String of' plants are two options that tolerate less light than other succulents

apartment into a welcoming home for a plant accustomed to the bright light and heat of South Africa or Central America? And how can we make more use of the available sunlight we do have?

Plant selection: Most succulents need at least 6 hours of bright light a day to thrive, but many are more forgiving than others when they receive less. I always recommend buying some varieties that will be easier to keep happy and close to true form in less than ideal light, even if you get grow lights. For a list of suggestions I really enjoy, see *'Chapter 9—Buying Guide: Plants - Part Sun.'*

Grow lights: We can't control the weather so there's no other way to be certain your plants will get enough light each day without artificial lights. This will be the main solution for those of you who live in cold locations that don't allow you to leave your plants outdoors, and I say embrace the lights and enjoy gardening in your home until the sun comes up. You'll revel in knowing you never need to worry about overcast weather. If you're located at a higher latitude that the sun doesn't quite reach year-round, you will become the owner of your very own sun. Grow lights are more energy efficient and affordable than ever so research what works for your space and budget. Set them up near a window to increase the available light and place your light-hungry succulents directly under the fixture while using the outer area for part sun plants. See *'Chapter 1—Basic Tips: Light – Grow Lights.'*

Cold frames, greenhouses: If you have an outdoor space that receives good light but it gets too cold or too much rain for your plants, a covered structure can turn that spot into a succulent haven. Options range from portable to semi-permanent and permanent shelters. Whether you plan to buy a prefab design or make it yourself, you'll need to do some research on how to add air circulation and additional warmth if needed.

Turn your pots: Finally, as I mentioned in Chapter 1, if your plants are getting decent natural light some days, you can turn them ever so often so the growth stays more balanced. In fact, this is a good idea for all of your plants that aren't growing under direct overhead light.

White walls help: White backgrounds will help reflect more light to your plants than darker colors.

Remember to cut back on watering and stop fertilizing when the light isn't right to help slow a succulent's growth until brighter days return. If your climate is succulent-friendly at certain times of the year, put your plants outdoors and let them get as much fresh air and bright natural light as they can tolerate, but remember to ease them into the sun to avoid damaging them, even when temperatures are mild *(see 'Chapter 3—Succulent SOS: Take Action – Increase Light')*. And if your plants are getting leggy or etiolated from inadequate light, see *'Chapter 3—Succulent SOS: Take Action – Etiolation Repair.'*

Hot—Humid

Examples: Gulf Coast states, South Atlantic states, Hawaii, Central Mexico, Uruguay, Zambia, Barcelona, Milan, Philippines, Northern India

Perks: Warm weather most of the year, sunny, possible mild winters

Challenges: Heat, humidity, overcast weather, rainfall, possible freezing in winter

Suggested genera: Kalanchoes, Aloes, Haworthias, Echeverias, Euphorbias, Stapelias, Cereus, Epiphyllums, Opuntias

More difficult genera: Sempervivums, Aeoniums, Crassulas, Lithops, Cotyledons

It's true that some regions just won't support certain varieties year-round without grow lights and other controls in place. Temperature extremes, long stretches of overcast weather, and too much precipitation are the primary foes of any gardener. Believe me, as a lifelong resident of Houston 'Bayou City' Texas, I know. When it's hot here, it's really hot—90°F (32°C) days start in May and don't cool down until September, and that's if we're lucky. Our humidity is thick enough to pour over pancakes and it rains and rains and rains. Some succulents don't mind the heat too much, but throw in high humidity and they start to freak out, so it's important we meet as many of their requirements as possible during these summer peaks.

To help combat the humidity, I've adapted my succulent potting mix to include as little organic material as possible, meaning very little 'dirt dirt' with the bulk of the mix being drainage materials. When I first started making my own mix, I used equal parts topsoil and drainage amendments but now I use a third-part topsoil or less.

Brown organic materials retain moisture, and the browner the soil, the longer it will take a pot to dry. Factor in a humid climate and there's a chance the soil won't dry out quick enough for your succulents. Drainage materials are your best friends and worth every effort to seek out and bring home, especially in hot and humid locations.

Also remember that the addition of any top

Ric Rac cactus and Euphorbia ritchiei (below) do well in heat and humidity

dressing can contribute to trapping excess moisture in your pots. Pumice and Turface are porous and some of the best options to help protect the lowest leaves from contact with damp soil without hindering evaporation.

Unless you have the light and space indoors, there will be days when you may have to protect your succulents from too much heat and direct sun. Air circulation is vital to moving the humidity away from your succulents and keeping them a bit cooler. I use an oscillating electric fan on my west-facing screened porch from May to August or later. I also added white curtain panels to filter the direct sun.

Remember that many varieties of succulents will go dormant during the hottest months. Aeoniums, Adeniums, and leafy Euphorbias (among other types) may drop their leaves during this time. You'll need to find a way to keep these plants cooler and reduce watering and fertilizing until the weather is milder again.

Dormant plants are very slow to uptake water and produce new growth, so while our instinct is to give them a drink more often when the temperatures rise, this can backfire and cause root rot and other problems.

I'm guessing if you're reading this, you have a lot of overcast weather too which means your succulents might start looking for more light. This is the primary reason I purchased my first grow light fixture. If you don't have artificial lights to help prevent etiolation, you'll want to discourage new growth until the skies are clear again. Since watering your plants encourages them to grow, be sure to consult the forecast prior to giving them a drink. If the week ahead looks cloudy for more than two or three days in a row, wait for sunnier days and definitely don't fertilize them.

See 'Chapter 9—Buying Guide: Plants - Part Sun' and 'Hot and Humid' for suggested varieties.

Air Plants are a fun addition to any plant collection

The arid gardens at the Barton Warnock Visitor Center in Terlingua, Texas

Hot—Arid

Examples: West Texas, Arizona, Nevada, Southern California, Chihuahua, Sonora, Luxor, Jaipur, Algeria, UAE, Alice Springs

Perks: Less risk of overwatering and rot, fewer pests and fungal problems, ample sunshine

Challenges: Heat, rainfall after long droughts, possible winter freezes

Suggested genera: Echeverias, Aeoniums, Cacti, Aloes, Agaves, Yuccas

Difficult genera: Crassulas, Haworthias, Gasterias, Euphorbias, Kalanchoes, Epiphyllums, variegated plants, small species or young plants

Succulents have evolved to survive in sunny, arid climates with little rain which is why they do so well in locations like California, Australia, and South Africa. Yet there are times when even these succulent utopias have to deal with heat waves, copious rainfall, and even freezing temperatures. In the last two years alone, California saw all three of these climate aberrations—record-breaking heat and rainfall followed by a late winter freeze.

I worried for my friends there who lovingly tended to their xeriscaped gardens. I'm well acquainted with what too much rain can do to succulents, especially when they're not used to so much hydration at once.

At first they were excited because their plants seemed to really enjoy the unexpected precipitation, but I knew when the sun came out they wouldn't be singing the same tune. Sure enough, I started hearing and seeing their reports of the damage and pleas for the rain to go away. Scorched and oversaturated succulents, pest outbreaks, and the freeze that following winter had my west coast friends struggling to keep their outdoor collections alive.

While succulents do love long, bright days more than most plants, there aren't many that can tolerate direct sun in high temperatures.

We tend to think of cacti as solar-worshipping desert dwellers, but even they try to find protection from the hot sun by growing behind rocks and other plants. They may seem to thrive under extremes, but the reality is they prefer a little shelter. Just like humans, most succulents are not going to reach their full potential if they're worrying about conserving energy to make it through a 100°F (38°C) day.

When the weather heats up in arid climates, you may need to provide shade to your succulents that receive direct afternoon sun. Shade cloth and full sun native plants can be used to filter and block direct light. An electric fan is a great addition if that's an option.

If it rains and the sun is still a few hours from setting, you might want to dry your more sensitive plants with a leaf blower, fan, or another method to keep them from scorching. Should your drought-accustomed succulents get too much rain, you can always dig up your favorites and repot them when the weather dries up again.

If a frost or freeze is imminent and you can't bring your plants indoors, stop watering and cover them with frost cloth. See the next section *'Regional Tips: Cold'* for protection methods.

Survivor: One lone Opuntia paddle still stands thanks to the shade it found from a cliffside Agave; Lost Pines Trail, Big Bend National Park, Texas

Cold

Examples: Temperatures below 60°F (15°C); locations that experience frost, freezing, snow

Perks: Colorful succulents, fewer pests, vernalization leads to more flowers

Challenges: Frozen or frost bit succulents, slower growth, soil retaining moisture too long

Suggested genera: Aeoniums, Agaves, Sempervivums, Yuccas, Stonecrop Sedums, cacti native to colder regions

Difficult genera: Echeverias, Aloes, Kalanchoes, Lithops, tropical succulents

While many succulents are known to be cold-hardy and frost tolerant, it's important to understand how the climate where you live influences their hardiness. If you order succulents from regions that don't get as cold as yours, they're not going to be as hardy as those procured from a grower in your area or a similar climate. And if you live somewhere that has mild winters with the occasional surprise freeze, your succulents are even less cold-hardy than those accustomed to cooler year-round temperatures.

A few varieties of succulents can handle freezing conditions but the odds decrease greatly if they're in wet soil. If a cold front is headed your way be sure to stop watering in time for the soil to dry before the temperatures fall below 60°F (15°C.) If the temperatures won't drop below freezing for more than a few hours and your soil is dry, it should be safe to leave your succulents outdoors as long as you can cover them with frost cloths or sheets. Some people use a space heater, heating pad, or strings of incandescent lights to add warmth.

In the northern hemisphere the coldest wind tends to come from the north so move your plants to another location if you're keeping them outside. If you live south of the equator your plants will receive the coldest winds from the south. Move the pots up against a wall of your house where it will be warmer than those left out in the open.

Most 'soft' succulents like Echeverias and tropical natives like Kalanchoes are not frost

Sempervivums (top) and Agave havardiana (below) are great cold-hardy options

tolerant, and only a few species can stand being left outdoors uncovered for more than a few hours when temperatures drop below 32°F (0°C.)

You'll need to be prepared to overwinter your succulents indoors. Be extra cautious about overwatering and not fertilizing during fall through winter because you don't want to encourage growth during this time.

Before you bring your plants indoors, inspect them for any pests or other critters that have taken up residence in your pots. Spiders, centipedes, millipedes, and earthworms are not likely to abandon their home but you can put the pots in bins which will help contain any escapees until you can release them outside.

If you live somewhere like Houston where a cold front will come through for a few days followed by a period of warmer weather, you can either leave your plants inside if you have good light or move them back outside. Remember when moving succulents outdoors after being kept inside, or covered with sheets for more than a couple of days, it's important to slowly increase the light they receive or else their leaves may get burned.

Since I keep most of my succulents outside, I watch the forecast like a hawk if a cold front is heading our way. The fronts are usually preceded or accompanied by heavy rain so I do my best to move my potted plants to shelter to keep the soil dry. During longer stretches of cold, rainy weather, I store most of my collection in my garage under grow lights which help provide an extra 5-10 degrees of heat. If temperatures fall below 40°F (5°C) I leave the lights on along with a small space heater. I found I need to plug my lights into a power strip with a surge protector to avoid tripping the circuit breaker when using the heater.

If the following week looks warm enough to move my succulents back outside, I start with those I've put in the house since I don't have good natural light indoors, especially any Echeverias and varieties that are quick to etiolate. I leave the others under the grow lights until spring since they're happy where they are, and ain't nobody got time to move that many plants over and over again, right?

You'll know pretty quickly if a succulent has suffered from frost or freeze. Frost causes dimpled spots, and leaves that have frozen will turn translucent and may even melt and drip. It's not a pretty sight but there is a chance the stem and roots will survive long enough to put out new growth. Cut back the damaged portions and keep the rest potted up. I thought I lost a Sansevieria to a freeze but thankfully it bounced back after several months.

Should you want to try keeping succulents outdoors year-round, opt for hardy varieties of Sempervivums and Stonecrop Sedums. Their shallow roots and thin leaves are less sensitive to frost and freezing temperatures but it's still important to keep them dry. Don't be afraid to experiment with other plants that originate from colder regions while keeping in mind that their hardiness decreases the longer they've been growing in warmer locations.

Mild Winters

Some regions experience very mild winters with little to no seasonal changes, or the cold weather isn't consistent. If you're keeping succulents native to locations with similar climates to yours, you can continue to encourage growth year-round. But if you have plants that are inherently winter dormant, it's recommended to allow them to experience a period of dormancy during their normal rest periods.

Mixed

I think most of us fall under this category—hot summers and cold winters with a chance of freezing and snow. Some of you have both dry and humid seasons to contend with. Please refer to each section that relates to your climate to get a better idea of the adjustments you can make throughout the year. Where weather 'mood swings' are common, you'll be better prepared to protect your outdoor succulents from the elements on the fly.

Examples

South Florida: Combination hot/humid, lack of sunlight
Maine: Combination cold/humid, lack of sunlight
Northern California: Combination hot/arid, cold
Gulf Coast: Combination hot/humid, lack of sunlight, cold

Indoors

When growing indoors, temperature extremes and precipitation aren't factors, but humidity can still be a concern. Without grow lights, you may not have enough natural light indoors to keep your succulents from stretching. Without a breeze, the leaves and stems of your plants may become weak. And without the natural elements to contend with, indoor succulents miss out on the chance to learn how to adapt to the flux of conditions outdoor life offers. There are some simple modifications you can make that will help your indoor plants thrive.

Wind: Adding a fan on a low speed to your setup will help regulate the humidity in your home and give your plants a workout. I've found this gentle resistance helps stems bulk up and teaches leaves to hang on tighter. Ceiling fans work well if you have already have one, as do oscillating standing fans.

Light: In the northern hemisphere, south and west facing windows offer the most light, while east windows get morning sun. North windows tend to receive the least light, so opt for the other directions. Turning your plants on occasion helps all the leaves get a chance to receive equal light. Remember glass reflects sunlight, and some windows are going to filter the natural light to a degree, especially if they are covered with screens. If you're still having trouble with etiolation, do consider treating yourself and your plants to a grow light.

Natural elements: You may have already experienced the difference in growing outdoors versus indoors and seen how well succulents respond to being outside. They seem to wake up and realize they've got more work to do. Maybe it's being around other plants, having more space around them, or feeling the natural sunlight and wind between their leaves, but something happens that encourages them to grow more robustly. Try to put your succulents outside when you can or open the windows to help them feel like they're outdoors.

Cautions

Windows: I've seen cases where succulents kept on a window sill suffered frost damage because the leaves were touching the glass during a cold snap. They can also burn through the glass, so be mindful of the temperature near your windows and move your plants if necessary.

Air vents: Placing plants too close to air conditioner and heating vents can stress them out.

Pests and fungi: Inspect any new plants thoroughly since they will be in close quarters indoors, and pests and fungi can quickly spread to the rest of your collection. Indoor plants are more prone to fungal gnats so it's important to let the soil stay dry for a day or more to discourage them from laying eggs in your pots.

Kids and pets: If you have kids or pets that may try tasting your succulents, be aware that many varieties are toxic to humans, animals, or both. It's very important to research any plants that will be within reach of children or pets, but it's also as simple as an internet search for the variety's name with the keyword 'toxicity.' The ASPCA website also has a great database on plants that are toxic to cats and dogs.

A few examples of animal-toxic succulents include Euphorbias, Kalanchoes, Aloe Vera, Sansevieria, and Portulaca. Pet-friendly options include Beaucarnea recurvata 'Ponytail Palm,' Burro's Tail Sedum, and most Haworthias and Echeverias.

Kids and pets are also notorious for damaging plants. Children may try to help you remove all the leaves from your succulents. *All* of them. And I'm sure my dogs aren't the only ones who enjoy curling up in propagation trays.

If you have cats, be prepared to find your plants on the ground if they're hogging the sunny spots in your home. Cats do not like sharing their sunny spots.

Tiger with Pilea peperomioides

Chapter 5
IDENTIFICATION

"Think for a moment: What are plants doing? What are plants all about? They serve human beings by being decorative, but what is it from its own point of view? It's using up air; it's using up energy. It's really not doing anything except being ornamental. And yet here's this whole vegetable world, cactus plants, trees, roses, tulips, and edible vegetables, like cabbages, celery, lettuce - they're all doing this dance. And what's it all about?"

—Alan Watts, The Essence of Alan Watts

Identification

Some people aren't as eager to learn and memorize the names of all of their succulents, while others can't fall asleep at night without a positive identification of a new plant. The latter also tend to keep detailed logs of the names, dates purchased, and notes on their growth. As long as you know the general care guidelines for your plants it's fine to just enjoy them anonymously but you may miss out on the education that comes with researching IDs.

I have a spreadsheet with several columns including the date and location I got them, common names, botanical names, geographic origin, links to the webpage I identified them by, notes on their growth, and a few other columns on care specifics. Obviously it got to be a bit tedious and ridiculous, so now I aim to just fill out the first seven. That may seem like a lot of data entry but it has been so helpful to refer back to, I always thank myself later.

Date	Source	Scientific name	Common Name/Notes
6/17/2003	Mom	Aloe barbadensis	Aloe Vera
11/1/2015	Joshua	Senecio radicans	String of Bananas
2/24/2016	HEB	Echeveria setosa	Echeveria Firecracker
3/31/2016	Joshua's Native Plants	Adromischus cristatus	Key Lime Pie
5/7/2017	Bob Smoleys	Ledebouria socialis	Silver Squill
5/13/2017	Local cactus sale	Edithcolea grandis	Persian Carpet Flower
4/6/2018	Swap 2 - Evelyn	Portulacaria afra	Trailing variety
9/26/2018	Kathy	Kalanchoe synsepala	Walking Kalanchoe
9/7/2019	HCSS	Mammillaria plumosa	White fluffy clumping
9/1/2020	New Roots	Consolea rubescens (Opuntia)	Road Kill Cactus
2/7/2021	East Austin	Peperomia axillaris	Taco Plant
3/26/2021	Jody	Cistanthe grandiflora	Rock Purslane/ Calandrinia

A glimpse of my personal spreadsheet for logging new plants

Research = Recognition

While there's no telling how much time I've spent hunting for IDs, I picked up a lot of knowledge along the way. It was thrilling when I finally started seeing other succulents and knew the names—usually because I had already ID'd one in my collection. Eventually it became easier to tell the difference in Aloe leaves versus Haworthia, or an Echeveria versus a Graptoveria. Soon, I could distinguish a West Coast grown Echeveria 'Perle von Nurnberg' from a Gulf Coast 'PVN.'

Spend enough time researching identifications and you'll notice something akin to intuition combines with your previously acquired learnings, allowing you to reach quicker answers with more confidence. You begin to develop a discerning eye for coloration, leaf-shape, growth patterns and other details that help close in on a possible identity. After narrowing down the options, you reach a manageable lineup of suspects and begin eliminating them one by one. On occasion you may have to wait until a succulent blooms to confirm an ID.

As with the E. Perle von Nurnberg, I mentioned the appearances of some varieties are quite different depending on where they're being grown, so you may be correct about an identification even if yours doesn't look the same as one in a comparison picture.

Botanical Names

Knowing the genus and species of your succulent helps you research its particular care needs more thoroughly. It will also aid in determining hybrid species. Some common crosses include:

Aloe x Haworthia = Alworthia
Echeveria x Sedum = Sedeveria
Echeveria x Graptopetalum = Graptoveria
Gasteria x Aloe = Gasteraloe
Pachyphytum x Echeveria = Pachyveria
Sedum x Graptopetalum = Graptosedum

Since many succulents share identical or very similar common names, knowing the botanical name (also called 'scientific,' 'binomial,' and 'Latin' name) gives you the advantage of having one precise ID to work with while training your brain to recognize the connections between other genera and species. This 'official' name consists of two names, the first being the genus name ('genera' plural) and the second the species.

'Hen and Chicks' is the common name used for Sempervivums and Echeverias like E. imbricata which offset at the base, just like a mother hen sitting on her brood.

If I said I have a 'Mother of Thousands,' that could mean a lot of different plants. You'd know it was probably a Kalanchoe but from there it could be K. pinnata, K. tubiflora, K. daigremontiana, K. fedtschenkoi...you get the idea. The commonality between all of these is their ability to create copious plantlets along their leaf margins which drop to the ground and mature to create their own plant horde, hence the very appropriate common name.

There's also some misinformation in common names. Euphorbia tirucalli is sometimes called 'Pencil Cactus' although it's not a cactus. 'Coral Cactus' is actually a Euphorbia lactea usually grafted on Euphorbia neriifolia. Beaucarnea recurvata 'Ponytail Palm' isn't a true palm tree although they can grow large enough to right-fully be called a tree.

Sedeveria 'Starburst'   *Graptoveria 'Debbi'*

Families

There are several families in the succulent taxonomic rank but the majority of identifications will fall under Aizoaceae, Asparagaceae, Asphodelaceae, Asteraceae, Cactaceae, Crassulaceae, and Euphorbiaceae.

Note: The 'ceae' at the end of the Latin names is pronounced 'see-eye' or 'see-ee.'

Aizoaceae includes Mesembs and 'ice plants' among others
Origin: South Africa
Examples: Pleiospilos nelii, Delosperma cooperi, Lithops salicola
Family: 190 genera with 1,900 species

Asphodelaceae includes Aloe, Gasteria, Haworthia
Origin: Temperate and tropic zones
Examples: Aloe humilis, Gasteria *'Little Warty,'* Haworthia pumila
Family: 40 genera with 900 species

Asparagaceae includes Yucca and Agave...and Asparagus, of course
Origin: Worldwide
Examples: Albuca spiralis, Agave attenuata, Sansevieria cylindrica
Family: 140 genera with 2,900 species

Asteraceae includes Senecio and Othonna along with 1000s of others
Origin: Worldwide
Examples: Senecio vitalis, Argyroxiphium virescens, Othonna arbuscula
Family: 1900+ genera with over 32,000 species (mostly non-succulents)

Cactaceae includes cacti
Origin: Native to the Americas
Examples: Echinocactus ingens, Mammillaria heyderi, Cereus peruvianus
Family: 140 genera with 1,500 species

Crassulaceae includes Crassula, Echeveria, Aeonium among others
Origin: Worldwide
Examples: Sedum morganianum, Dudleya farinosa, Kalanchoe synsepala
Family: 34 genera with 1,500 species

Euphorbiaceae includes Euphorbia and Jatropha among others
Origin: The tropics of Indomalaya, Americas, and Africa
Examples: Jatropha berlandieri, Euphorbia trigona, and the holiday Poinsettia
Family: 300 genera with 7,500 species

Resources

As far as resources go, the internet is my main tool for finding identifications. Along with educational websites, I have found online succulent shops helpful as well. I also utilize social media platforms to compare my plants with pictures of others. The more photos that are hashtagged with a plant's name, the better I feel about my identification. Knowing how to phrase a search 'question' is a huge boon to getting closer to an answer.

You'll find more anatomical information and useful terms to help you with identification in the *'Knowledge Bank.'*

Basic Botanical Anatomy

Stem: The main supporting stalk of a plant and any branches; use the length, thickness and other traits to identify a plant.

Rosette: The arrangement of overlapping leaves in a compact spiraling or circular form. Echeverias, Haworthias, and Aloes all form rosettes though they look rather different from each other.

Leaves: The shape, texture, quantity, and growth pattern of leaves can help determine a genus and species.

Inflorescence: The whole structure of the flowering portion of a plant.

Flowers: The colors and structure of individual flowers and the arrangement of multiple flowers on a single inflorescence are helpful to finding the name of a plant. Bud, petal, style, stamens, and stigma colors are useful to identification. The style is the center 'stalk' of a flower that supports the stigma. Stamens are the pollen-producing organ surrounding the style. The pericarpel is the 'cup' that holds the flower.

Blooming season: Periods of the year when a plant flowers.

Ribs: In cacti, ribs are the accordian-like pleats along the main stem; the number of ribs helps with identification.

Areoles: The spots along cacti ribs that produce spines and glochids; typically where cacti flowers emerge; areoles are exclusive to cacti.

Spines: Needle-like structure radiating from cacti areoles. The length, color, and quantity aid in identification.

Glochids: Hair-like spines found in the areoles of cacti.

Cereus silvestrii by J.N. Fitch

Huernia barbata by S. Edwards

Lithops fulviceps by J.N. Fitch

With those terms in mind, we'll look briefly at how they're used in descriptions of Echeveria purpusorum and Cereus peruvianus.

Echeveria purpusorum has a short **stem** and its pointed **leaves** form a **rosette** with a mottled coloration. Its **inflorescence** is 7-11" (20-30cm) and the reddish-orange **flower** buds open to reveal yellow **petals.** The **style** is green and the **stigmas** are maroon.

Cereus peruvianus **stems** are green to grayish blue, **cylindrical** with 9-10 **ribs**; small **areoles** are widely spaced apart with a varying number of **spines**; spines are sometimes absent.

While those are paraphrases of the more elaborate physical descriptions of each plant, those are the main terms you'll be using to identify your succulents. See the *Knowledge Bank* for a glossary and more resources for educating yourself about botanical descriptions and definitions.

Echeveria purpusorum 'Dionysos'

Cereus peruvianus

Example 1: Kalanchoe humilis

Example 2: Echeveria pulidonis

Identification Online

When looking for an ID on the internet, you'll begin with search terms that describe your plant. Let's explore some examples of how to narrow down the options to a few possible answers. Remember that putting a word in quotes forces that term to be returned in the results, and adding a dash before the word will omit it from the results. You'll be using the image results as comparison photos.

Ex. 1- Search the terms: Succulent identification green leaves purple markings.

Now look at the image results. You'll see several varieties that apply to the search, but the plant we're looking for isn't an Echeveria. To omit 'Echeveria' add '-Echeveria' to your search. Now we're looking at a handful non-Echeverias with purple markings: Kalanchoe marmorata, Kalanchoe humilis, Ledebouria socialis, and a few others. I think the Kalanchoes both look similar to the plant I'm looking for, so I need to figure out which one it is. My plant has more of a striped pattern than the splotchy K. marmorata so I'm going to dig deeper into the Kalanchoe humilis by looking it up. Sure enough, the results match my succulent perfectly, and I can be confident I found the right identification.

Ex. 2- Search the terms: Succulent identification rosette green leaves red edge.

Looking at the image results, we have several options that don't look anything like our specimen and a few that resemble it very closely but the leaf shape and coloration is off. Our subject has red along its leaf edges, not just the tips. The outer edge of a leaf is called the 'margin' so let's edit our search to replace 'edge' with 'margin.' Now we're getting somewhere. It's not an Echeveria agavoides or Aeonium 'Kiwi' as our leaves aren't as pointed. It's not an Echeveria 'Dondo' or 'Tippy' either because their red features are more focused on the tips. It's not Echeveria nodulosa because its mid-leaf is lined with red as well. Wait, what's this? Echeveria pulidonis seems to fit the description but let's look it up on its own. The results for E. pulidonis return a lovely light green rosette with the leaf edges traced with a red outline. The intensity of the red varies as we discussed in 'Chapter 1—Basic Tips: Color and Form,' but it's quite apparent from the elegant leaf-shape that we've found the identity of our plant.

Visual Clues

Hybrids: Keep in mind that crossbreeds will usually carry more traits from one parent than the other. Just like human twins, hybrids of the same parentage may look noticeably different. For example, the leaves may be longer or more tapered, or thicker and rounder than another from same variety.

Variation: Some succulents are so highly variable they may look like a different species or a hybrid. This is common in Aloes and cacti.

Euphorbias vs cacti: Some Euphorbia species resemble a cactus. Both can have similar bodies and spines that radiate from conical raised points call tubercles.

One clue is to look at the point where the spine originates from the body. Cacti spines sprout from the center of a defined, often round and woolly radius called an areole, while Euphorbias don't have areoles. Another clue is the milky sap Euphorbias ooze when poked or cut. It's an irritant so be careful to wash it off your skin if you ever get any on you.

Flowers: Some of the smoother leaved varieties of Haworthia strongly resemble Aloe, but their flowers are quite different. The flower stalks of Haworthias tend be rather long, thin, and curving. The blooms are small, white, and spaced apart almost like a tiny orchid. Aloe flower stalks shoot straight up and they tend to be thicker. The blooms are larger, more colorful, and typically clustered together towards the tip of the spike. The mature flowers are pendulous, meaning they hang down like little bells.

Fuzz: *Tomentose* is one definition of a plant that is fuzzy. Because of this, many fuzzy succulents are referred to in their species name as 'tomentosa,' such as Kalanchoe tomentosa *'Panda Plant,'* Plectranthus tomentosa *'Vick's Plant,'* Cotyledon tomentosa *'Bear Paws,'* and Crassula tomentosa.

Conditions matter: Remember that your growing conditions affect the appearance of your succulent. Comparing it to others that are grown under different circumstances can be tricky to confirm. Do your best to narrow the options down to a few IDs and wait for flowers or colors to appear in future seasons. If your subject isn't growing true to form due to insufficient light, make the necessary corrections and try to ID it at a later date.

Keep in mind that one plant can have several botanical names known as synonyms, but there is usually one that is the scientifically accepted name. Sometimes a genus is split into two or more distinctive genera, while others may be combined into one, as with the merging of Notocactus and Parodia into Parodia alone. Even so, many people continue to use the former names.

The genus Bryophyllum was merged under Kalanchoe but some botanists are still adamant they deserve their own genus. Kalanchoe flowers are upright while Bryophyllums' are pendant. And it's Bryophyllums that produce plantlets along their leaves as with Mother of Thousands, while Kalanchoes propagate via offsets, yet you'll still find 'Kalanchoe' as the most commonly used name for these plants.

Now that you're familiar with how I look for IDs, you can try it on your own mystery plants. Once you feel good about an answer, you can enter it into your log, or for further confirmation, search for an active online succulent identification forum and upload a picture to get feedback from others.

Chapter 6
GENUS TIPS

"Careful! That'll poke your eye out."

—My great-granny Nadine Gayler on the Yuccas along her porch step, 1985

Aloe

With over 500 species, many varieties of Aloes are very easy going and forgiving as long as the climate is right.

Winter: Since most Aloes originate from warm and/or tropical regions, they're not very cold-hardy. They will suffer from frost damage and be among the first succulents to freeze if left unprotected.

Summer: Aloes like part sun and some can take quite a bit of direct sun if it's not terribly hot outside. They may go summer dormant in high temperatures and should be watered less frequently while giving them a more shade during this time.

Propagation: All Aloes flower, sometimes prolifically, and since they're attractive to pollinators, there's a good chance you will get seeds if the variety is self-fertile. They also propagate by offsets which quickly establish as independent plants.

Regional Tips: Enjoy them in pots, or plant them in the ground as long as you can offer protection from the elements, whether too much sun or when temperatures drop below 40°F (4°C.) Some varieties are hardier to colder weather than others as long as they are kept very dry. Remember to add extra drainage materials to your soil if you live in a humid region.

Aloe brevifolia

Gasteraloe cv. Royal Highness

Parodia (Notocactus) buiningii flowers

Cactus

With about 140 genera and over 1,500 species, cacti are held close but carefully to the hearts of many succulent lovers. If you don't already love them too, just wait. Before long, they'll start beckoning to you with their lovely geometric patterns and audaciously bright flowers, and one day you'll wonder how you weren't a big fan in the first place. This happens with other 'weird' genera too, and soon you may even like the strange varieties more than the pretty pastel rosettes.

Winter: Many cacti are naturally winter dormant. Some genera that originate from colder regions can tolerate freezing temperatures but for those that aren't cold-hardy, their soil should be kept bone dry throughout winter except for a small drink of water targeted at the roots every other month or so. This is a safe estimate you may need to adjust to a particular type, but you just want to water enough to keep the roots alive. If you keep them in the ground or outdoors, you will need to provide protection from frost and freezing temperatures.

Summer: Cacti can go dormant during the peak of hot summers. It's best to water a bit less deeply and reduce the frequency during this time.

Propagation: A cactus will flower when it is mature enough and when the right boxes are checked for its particular needs to bloom. Bright sun and a period of vernalization are usually required but some cacti flower eagerly and regularly without fuss. Some need cross-pollination to set seed and some, like Opuntias, produce delicious fruits. They offset through clumps, branching, and developing new pads and sections.

Regional Tips: Providing adequate light is the main challenge to keeping cacti in good form, especially when fertilizing. If they are actively growing but not given enough light, they will quickly exhibit etiolation which is harder (or impossible) to repair in many forms of cacti.

Mammillaria elegans fruits containing seeds.
Many Mammillarias produce these tasty 'chilitos,' and while they look like tiny chilis, they're not spicy but tangy and sweet.

Gymnocalycium 'Moon Cactus'

Domino cacti flowers are often beautifully fragrant (Echinopsis subdenudata)

Echeveria 'Lola'

Echeveria

The genus Echeveria is named after the 18th-century Mexican botanical artist and naturalist Atanasio Echeverría y Godoy. There are about 150 species with countless varieties and hybrids.

Winter: Echeverias tend to be opportunistic growers but depending on their parentage, they may need to be overwintered. They and their hybrids are rarely frost-tolerant and because of their succulence, they are among the first to freeze. Water with a light hand during winter to discourage growth while the daylight hours are sparse.

Summer: Some species of Echeverias are less eager to grow in high temperatures and most will slow their growth or die in too much hot direct sun. Morning to early afternoon sun with part sun in the afternoon will keep most Echeverias happy year-round.

Propagation: Most Echeverias propagate by leaves and cuttings. They also produce offsets at the base and along the stem, and some varieties will produce flowers that grow babies on the stalk, all which can be established as individual plants.

Regional Tips: Echeverias thrive in cool to warm and dry climates. They struggle when temperatures are too hot or cold and will either go dormant or die if left unprotected from extreme elements. Experiment with different varieties to find what works best for your growing situation.

Macro look at the flower and nectar of an Echeveria 'Lola'

'Shattering' Echeveria flower bud (E. diffractens)

A very large Echeveria 'Topsy Turvy' and flower stalk

Euphorbia lactea tree in Jamaica

Euphorbia

Euphorbia boasts over 2,000 species in its incredibly diverse genus. Many are similar in appearance to cacti while others resemble trees and shrubs. Not all are succulent, but most species produce a toxic milky white latex when cut, so handle with care. Since this genus originates from all over the world, the growing seasons vary. You'll notice Euphorbia flowers are rather unique in that they're actually a cyathium or a cluster of smaller flowers surrounded by modified leaves that can resemble petals called 'bracts.' This feature is unique to the genus. Some are deciduous and will naturally drop their leaves in winter or summer.

Winter: While many Euphorbias are cold-hardy, particularly the leafier 'spurges' like E. myrsinites, the more succulent varieties like E. trigona and E. tirucalli 'Sticks on Fire' are frost tender and will die in a freeze if left uncovered outdoors. Increase the odds of survival by keeping their soil dry.

Summer: Euphorbias are usually opportunistic growers and if the climate is temperate most growth will occur between spring and late fall. In hotter climates they may need to be treated as summer dormant.

Propagation: Euphorbias flower and produce a capsule-like fruit containing seeds. Once it dries, it bursts open and flings the seeds far and wide which is why growers wrap the pods in a piece of cloth to contain them. Many branching varieties can be propagated by cuttings such as Euphorbia milii 'Crown of Thorns,' and others by removing offsets, like Euphorbia obesa.

Regional Tips: Due to the vast number of species originating from myriad climes, you'll find plenty of Euphorbia types that will thrive in your location. Most enjoy a lot of sun and heat, and the tropical natives want more water than others from drier regions.

Euphorbia trigona 'Rubra'

Euphorbia suzannae

Euphorbia obesa

Haworthia arachnoidea

Haworthia

The Haworthia genus has about 60 species and numerous hybrids, ranging from inexpensive to hundreds of dollars for the rarest varieties. Some are fenestrated, meaning all or a portion of their leaves are translucent and allow light to shine through. Most are on the small side and do well in part sun while some appreciate a lot of bright indirect light. They enjoy a deeper pot as they tend to put out long roots. Many Haworthias absorb some of their old roots annually for nutrition.

Winter: Haworthias are not cold-hardy and need to be protected from frost and freezing temperatures. Avoid overwatering during cold months when they are less actively growing to prevent root rot.

Summer: Haworthias can go summer dormant during the hotter months and need less watering during this time.

Propagation: Haworthias naturally propagate by offsets, seeds and some may produce bulbils on the flower stalk. Certain varieties are easier to propagate by leaf than others, and the roots of several types with larger tuberous roots can be used to grow a new plant. Topped Haworthias can also produce new individual plants.

Regional Tips: In arid climates, Haworthias may need more frequent watering. For humid climates, be sure to let the soil dry between waterings to avoid root rot. When grown indoors, they'll need as much natural light as you can provide. If you're using grow lights, set them to run about 14 hours a day.

Haworthia flowers

*Alworthia 'Black Gem';
a hybrid of Aloe speciosa and
Haworthia cymbiformis*

Haworthia coarctata

Anole keeping watch under the grow lights

Lithops

Lithops (*'Lithops' singular*) are easily one of Earth's oddest plants making them highly popular with succulent lovers. They're also infamous for being easy to kill. In fact, I bet we'd be hard pressed to find someone who has never killed a Lithops unless they've just never had one before, and that doesn't count. I know I took at least three victims before I really 'got' them and learned to respect how little they needed me to survive.

Due to our vastly different climates and growing situations, whether indoors or out, arid or tropical, and everything in between, it's virtually impossible to give universal advice on general succulent care, let alone Lithops. Would you believe me if I told you watering them once a month can be too much? It's true, and it can mean the death of a Lithops if watered at the wrong time.

But these little weirdos are great at expressing their needs and once we understand their body language and growth cycle it becomes much easier and less stressful to keep them happy.

Natives of the driest areas of South Africa, they live their low-key lives (or should I say low-leaved) among plains and outcrops where they mimic the rocks common to their location to hide from thirsty critters. And while they're from the southern hemisphere, most Lithops adapt to the growing seasons where they're being cultivated.

ithops salicola: The lightly fragrant flowers emerge late summer through fall and bloom for several days

Growth Cycle

Once you understand their growing patterns, it's easier to accept why they need so little water. Let's start with their flowering phase. Most Lithops need to be at least 3 years old before they'll flower. They usually bloom sometime between late summer through fall. After their flowers fade, they begin growing a new plant within the outer leaves, but you can't see it yet.

Through winter and into early spring, the new plant continues to grow while the outer leaves begin to wrinkle and shrink. The new leaves subsist solely on the water and nutrients from the old leaves, and for this time the root system is basically put out of service.

When the new growth becomes large enough, the outer leaves begin to split and dry out until the new plant fully emerges. Roots that dried out are replaced by new roots. Depending on the climate, Lithops in hot summer settings may go dormant or partially dormant until it's time to flower again. This cycle repeats each year, with new growth bursting through the old. Knowing all of this, when you buy a new Lithops and it looks like it's not flowering or splitting, if you still aren't sure of which growth stage it's in, you can look to the season for a clue.

Light

Now let's get into the care needs for these oddballs. As with other succulents, the most common causes of a Lithops' demise are overwatering and inadequate light.

Lithops salicola in spring: Old growth vs new

In nature, Lithops have adapted to their harsh conditions by growing with only the very top surface visible above ground. The light needs to be bright in order to reach the chlorophyll safely stored deep down inside the subterranean leaves. While most people don't pot their Lithops as deeply as they grow in the wild, they will still need at least 3 to 5 hours of bright light, preferably gentle but direct, and as many more hours of bright indirect light as you can provide—ideally 12 to 16 hours.

If you don't own a grow light, you'll want to find the brightest spot you can, away from any rain but also protected from full sun when it's hot outside. Remember, morning sun is gentler than afternoon, so east-facing windows and patios are ideal when the light isn't blocked by other objects.

Lithops etiolate and grow taller when they're not getting enough light. If this happens to yours, gradually extend its exposure to more light so it can photosynthesize enough to produce a new plant and keep the next generation true to form.

Before we talk about watering Lithops, let's cover their container and soil requirements.

Pots

For such small plants, Lithops can put down some pretty long roots, so it's important to pot them in a container deep enough to accommodate them. Clay or plastic pots both work as long as they have ample drainage holes and the planting medium is very fast draining. I do recommend clay pots for those new to Lithops as a precaution against moisture retention. I use both.

Lithops karasmontana

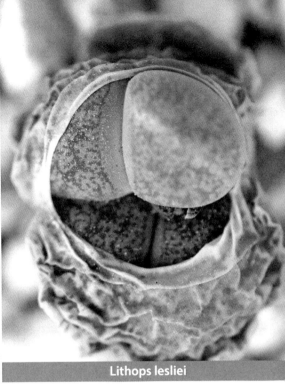
Lithops lesliei

Soil

As for soil, Lithops really need a potting medium that dries within 3 days or fewer. Most commercial plant retailers sell Lithops in the same soil used for non-succulent plants, but this doesn't mean we should leave them in that dark organic mix. I highly suggest having a fast draining potting mix ready before purchasing any Lithops, and the less organic matter in the mix, the better. I use about 90% non-organic materials to topsoil. My climate is terribly humid most of the year and it takes longer for soil to dry than Lithops prefer.

So if you're in a humid location, please believe me when I say investing in a bag of good drainage amendment is probably the best and easiest way to keep your Lithops alive and reproducing. There are alternatives to pumice, like expanded shale, Turface, scoria, and bonsai soil—even a combination of perlite with a bit of topsoil is better than regular potting soil.

Not only does potting in non-organic materials reduce the odds of overwatering or moisture retention, it helps prevent fungi and bacteria and makes the pot virtu-ally uninhabitable to pests like fungus gnats and root mealybugs. If you live in an arid region, or grow indoors with the required light, and you can master the growing cycle of Lithops well enough to know when to avoid watering, then you can get away with more organic material in your mix.

It's common to see a lot of Lithops planted together in one container. It looks fantastic but this potting situation can become problematic if the Lithops are at different stages of growth where one is in need of water but its neighbor is fully hydrated. I'd wait until you're comfortably familiar with the different species and growth cycles before putting them in the same pot, then go for it!

Ventilation is also important in stiflingly hot climates, especially when humid. By late spring, I've already activated my oscillating fan on my porch. When I start to feel like I'm slowing down from the heat, I assume my plants are feeling it too.

Water

We covered everything else first because without the right light, soil, and container, any amount of water at any time can kill them. Watering at the wrong time in their growth cycle can be the kiss of death for a Lithops, but the odds decrease if everything else is in place.

The best way to tell if your Lithops need water during the time when it's okay to water is by observing them. They'll start wrinkling or puckering, or maybe even appear to be sinking deeper into the pot. If you give them a light squeeze, they feel softer than when hydrated.

The tricky part about all of this is they do the same thing when they're about to shed their old leaves to allow the new growth to come in. That's why it's so important to know what stage of growth they're in before you water them. All you really need to remember is to only water after the old outer leaves are dry and stop watering after the flower begins to die (if it's mature enough to flower). Flowering typically occurs between late summer and the end of fall. New growth occurs during fall and spring, and old leaves dry out between late spring and early to mid-summer. Those are all wide open estimates, but a good guide nonetheless.

The main reason you shouldn't water after flowering or while new growth is forming comes down to the way Lithops utilize water. As I mentioned, the old leaves are the source of nutrition and water for the new plant that forms within. The root system is on pause for this time. If you water them during this phase, you risk confusing the plant into using water from the roots while it is actively absorbing the old leaves, which can engorge the plant beyond repair. You also risk root rot since the root system's activity is suspended and the excess moisture surrounds the plant with nowhere to go.

After the old leaves have dried up, you can give your Lithops a deep watering. This will probably be around late spring to mid-summer but the timing can vary. Put the pot in a saucer and slowly give it a good drink. Wait until it runs through the drainage holes. Dump the saucer. That's it! It can take a couple of days to plump up and look hydrated again. Water it again if it shows signs of wrinkling after 3-4 weeks but only enough to run through the bottom once.

If a month has passed without your Lithops showing signs of wrinkling during the summer months and you haven't watered them, you can moisten the top layer of the pot to help give the roots some moisture. There's a chance it is becoming dormant from the heat or natural cycle, so too much water can cause it to swell and split...and die.

Giving Lithops the right amount of water during the right time will sustain it through its flowering, fruiting, and new growth cycles. Some can go 6 months or more without a drink. Even in the driest climates, watering once or twice a month (during the right time) at most is the norm. If you can respect that fact about these plants, and you can give it enough bright light, then I know you can keep them alive.

Lithops Watering Guide

Summer: Water once the outer leaves have dried out and shed, and only if the new leaves look rather wrinkled. Water once and wait until the pot dries to see if the Lithops needs more water. A few wrinkles are perfectly fine. If it's humid in your region, the old leaves may not completely dry after the new growth fully emerges. You can use small sterilized scissors to remove them.

Fall: You can water as needed through the flowering period. If your Lithops doesn't flower, it's probably still too young. It's best to allow the flower to dry naturally then gently pull it from the plant.

Winter: Lithops need to be kept warm and dry during the winter season. They don't need water at this time since the new inner growth is absorbing the old outer leaves.

Spring: By this time your Lithops is probably looking quite rough and shriveled. Resist the urge to water as the new growth emerges. Over the next few months, the new leaves will continue to absorb the old. It's important to wait until the new growth is showing signs of wrinkling before watering. If your Lithops struggles to absorb the outer leaves, gradually give them more light.

Seedlings and younger plants: From my experience, Lithops seedlings need more frequent watering until they are at least a year old. Some may split as early as 6 months after germination. I continue watering when the soil dries rather than waiting for them to shrivel.

Lithops should split from the top...   *...but sometimes they split from the sides.*

Mesembs

There are many types of Mesembs that require a different care regime than Lithops. While they're grouped under the common names 'Split Rock,' 'Mimicry plant,' and 'Living Stones,' unlike Lithops, many Mesemb varieties like more frequent watering or have opposite growing seasons. Some don't absorb their old leaves as Lithops do. It's important to know what kind you have in order to care for it properly.

Most Mesembs require several hours of bright indirect light along with a few hours of gentle direct light if it's not too hot. They need very little organic material in their soil and tend to go heat dormant. Some will retain two or more stacks of leaf-pairs while others need to be allowed to absorb their older leaves.

Names & Examples	Absorbs Leaves	Water Needs	Growing Season	Bloomir Seasor
Aloinopsis *luckhoffii, rubrolineata, schooneesii*	X	Year round. Reduce summer and winter.	Spring and fall	Late wint
Argyroderma *aureum, brevipes, delaetii, framesii, testiculare*	✓	When shriveling spring through summer. Reduce or stop fall through early spring.	Fall through mid-winter	Fall through mid-wint
Cheiridopsis *candidissima, derenbergiana, herrei*	✓	When shriveling spring through summer. Reduce or stop fall through early spring.	Late winter through spring	Fall and spring
Conophytum *bilobum, calculus, flavum, jucun-dum, obcordellum*	✓	When shriveling spring through summer. Reduce or stop fall through early spring.	Fall through mid-winter	Late summer through mid-wint
Dinteranthus *inexpectatus, microspermus, wilmotianus*	✓	When shriveling spring through summer. Reduce or stop fall through early spring.	Spring through summer	Late summer through mid-wint

For instance, if Pleiospilos nelii has more than 2 pairs of leaves it will be at risk for rot due to the plant retaining too much water for its physical structure and root system. They should go unwatered until the oldest (third) pair is absorbed.

Each genus has different needs and growth cycles so be sure to research each type and factor your climate into your care routine.

Names & Examples	Absorbs Leaves	Water Needs	Growing Season	Blooming Season
aucaria lbidens, candida, elina, tigrina	X	When shriveling mid-summer through early winter. Reduce or stop spring through summer.	Fall through winter	Late summer through early-winter
enestraria hopalophylla 'Baby Toes,' hopalophylla urantiaca	X	Moderately, late fall and early spring. Reduce or stop summer and winter.	Winter	Late fall and early spring
apidaria nargaretae monotypic enus)	✓ 6 pairs max	When shriveling mid-summer through early winter. Reduce or stop spring through summer.	Fall through winter	Late summer through early-winter
leiospilos ompactus, nelii, elii cv. 'Royal lush,' nobilis	✓ 2-3 pairs max	Weekly, late summer through early fall. Stop during winter. Light watering spring.	Late fall and early spring	Late summer through fall
itanopsis alcarea, ugo-schlechteri, rimosii, chwantesii	X	Moderately, late fall and early spring. Reduce or stop summer and winter.	Late fall and early spring	Winter

Other Popular Varieties

Aeoniums

Native to the Canary Islands where the Mediterranean climate is mostly mild year round (64-75°F/18-24°C). You'll see the most growth from winter through spring in temperate regions. Most species are monocarpic.

Winter: Aeoniums can be fairly cold-hardy if the soil is dry but may enter dormancy in temperatures lower than 55°F (12°C.) Watering frequency should be reduced to once a month or less. They should be protected from frost and freezing.

Summer: Since Aeoniums tend to go dormant during the hottest times of the year this means you'll need to provide them a break from the heat by moving them to a part sun location or a bright spot indoors, and cut back on watering to avoid stem and root rot. The rosette may close up and some lower leaves could dry out and drop.

Propagation: Seeds and cuttings; some thicker leaved varieties may propagate by leaf.

Regional Tips: A combination of humidity and heat are the main challenges to keeping Aeoniums happy. I've had the most luck when I bring mine indoors for the summer when the days are regularly above 90°F (32°C.)

Crassula 'Jade plants'

Crassula ovata is commonly called 'Jade' and there are several varieties with unique leaves and colors like 'Hummel's Sunset,' 'Gollum,' and 'Hobbit.' Portulacaria afra 'Elephant Bush' is also referred to as Jade and has similar care needs. They prefer bright indirect light but can tolerate a few hours of gentle direct morning sun if it's not terribly hot outside.

Aeonium cuttings

Crassula 'Ogre Ears' leaf propagati

'Ogre Ears' with mineral deposits

Winter: If the soil is kept dry, they can be cold-hardy (25°F/-4°C) but it's best to protect Jades from frost and freezing temperatures, particularly in humid climates.

Summer: Avoid direct sun and overwatering as Jades are great at storing water in their leaves and stems. They may go dormant in high heat.

Propagation: Seeds, leaves, cuttings.

Regional Tips: Jade plants are some of the easiest succulents to grow since they do well indoors where the climate is controlled. If there's a chance of frost or freeze in your region, simply move them to a bright location indoors. Humidity can encourage powdery mildew so air circulation is important. If you see white spots on their leaves, they're most likely mineral deposits from the water. Wipe with a warm damp cloth and use filtered water to flush their system on occasion.

Kalanchoes

Kalanchoes and Bryophyllums are similar genera that are usually both grouped under Kalanchoe. Most are tropical African natives and are among my very favorite types of succulents. They're easy growers that don't mind the sultry summers in Houston and come in hundreds of collectible varieties. A few species are monocarpic as with Kalanchoe *'Donkey Ears'* and *'Flapjacks.'*

Winter: With origins from warm climates, these plants must be protected from frost and freeze. They can tolerate temperatures into the high 30s (+2°C) but their soil should kept dry. Watering frequency should be reduced to once a month or so to prevent root rot.

Summer: Kalanchoes can tolerate high temperatures as long as they're not in full sun. They're fine with a bit more organic material in their soil and most can handle more frequent watering than other succulents.

Kalanchoe laetivirens leaf plantlets

K. millotii and beharensis minima

Propagation: Seeds, leaves, plantlets, cuttings.

Regional Tips: Since Kalanchoes like a lot of light, they're happiest when grown outdoors and prefer warmer climates. When kept indoors they should be encouraged to grow slowly to avoid etiolation. This means no fertilizer, low watering, and using a smaller pot which tends to keep their size down. For those that are naturally tall, a fan should be kept nearby to help strengthen their stems.

Sedums

Referring to the creeping 'Stonecrop' varieties of Sedums such as S. angelina, S. reflexum, and S. dasyphyllum; most are native to temperate and subtropical regions around the world. As lithophytes (plants that grow in or on rocks) they grow best when offered a rocky surface to attach to.

Winter: Many types are cold-hardy and can even tolerate freezing temperatures if their soil is kept dry.

Summer: Stonecrop Sedums are less tolerant to heat and should be protected from direct sun during the hottest months of the year.

Propagation: Sedums easily propagate by cuttings and division. Wait until a new plant is established before breaking up a clump or else it can suffer from shock.

Regional Tips: Depending on the origin of the species, some Sedums are better at handling hotter climates than others. Research the type you're keeping to get a better understanding of its preference.

Sempervivums

Sempervivum may mean 'always living' but they are mostly monocarpic. They originate from a wide area ranging from the Alps to Iran. As with Stonecrop Sedums, they too are

Sempervivum 'Oddity'

Sedum dasyphyllum

lithophytes and like a rocky surface to grow on.

Winter: Species indigenous to colder climates can tolerate the lowest temperatures if kept dry.

Summer: Most Sempervivums are not very heat tolerant and should be kept in part sun, particularly in humid climates.

Propagation: Seeds; division of the offsets.

Regional Tips: Similar to Aeoniums, 'Semps' are less heat tolerant in humid climates and may enter a rest period. Reduce watering and avoid direct sun until the weather is mild again.

Senecio 'String of...' plants

'String of Pearls,' 'Raindrops,' 'Tears,' 'Bananas,' 'Dolphins,' 'Fishhooks'—these South African natives are lovely hanging plants that enjoy bright but indirect sun and mild to warm temperatures. When repotting Pearls, do your best to keep the root ball together as separating it can send the plant into shock.

String of Raindrops

Winter: Protect from frost and reduce or stop watering in temperatures under 50°F (10°C.)

Summer: Avoid direct sun during the hottest months and water every other week or so as long as the soil is drying out within a week.

Propagation: Seeds, cuttings and division; with cuttings, place the strand on soil and it will put down roots and form a tuber. You can also dig up tubers and repot them. It is rare for them to propagate via leaves.

Regional Tips: Besides Pearls, most varieties are fine in humid regions but will slow their growth in temperatures above 85°F (30°C) and below 50°F (10°C.)

It's important to let the soil dry for a couple of days between waterings during these periods to prevent root rot.

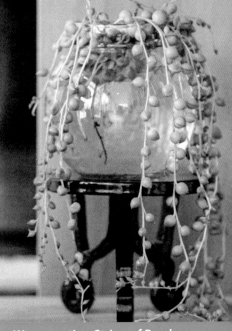
Water rooting String of Pearls

Chapter 7
IN-GROUND
SUCCULENTS

"For the last 40 years of my life I have broken my back, my fingernails, and sometimes my heart, in the practical pursuit of my favourite occupation."

—Vita Sackville-West, from a 1957 Observer column; author, poet, landscape artist

In-ground Succulents

If you're lucky enough to have a garden bed and you live in a climate with mostly mild winters, you should definitely try planting some of your hardier succulents in the ground. Although there will certainly be challenges like pests, rain, heat, humidity, and the occasional freeze or snowfall, a well-planned space will help protect your plants from the elements. You'll find that some succulents prefer a place in the garden, while others want to be kept in pots under more controlled conditions. Experiment with different varieties but be ready to dig them up and repot any that seem to struggle too much in the ground.

As I mentioned in *'Regional Tips—Cold,'* a bed built against the wall of your home receives more protection from rain and cold wind and also makes frost cloths more effective compared to a freestanding bed. North-facing beds aren't ideal as they're hit the hardest when a cold front moves in and usually receive the least sunlight. Remember east-facing beds get gentler morning sun, west-facing get hotter afternoon sun, and south-facing gardens get a little bit of both. The opposite is true in the southern hemisphere where the north receives more light and the south side will be colder. I'll walk you through how I built my west-facing succulent rock garden along with the modifications I had to make to protect my in-ground plants from rain, sun, and winter woes.

I started the first section of my succulent garden after a nice lady gifted me some really large Opuntia paddles from her xeriscaped front yard. I didn't have any pots large enough to hold them so I did what I could to amend the soil and stuck them in the ground. Two years and two winters later, they were still happy where I planted them so I decided to expand the bed by another three feet (1 m), then another. You may want to start with a smaller test section as well.

Borders

You'll need to build a border that allows the rain to run down and out of the bed. I used concrete garden edgers for the front boundary and stones for the sides while leaving small gaps between them to let water run through.

Soil

The soil in my bed had too much clay to allow moisture to evaporate quickly enough for succulents, and you'll probably need to modify your soil as well. I started by removing the first 6 inches (15 cm) and set it aside in a large bin where I mixed in topsoil, pumice, porous ceramic soil conditioner, and expanded shale. You'll need enough of the mix to fill the bed to the top of the border and even more to build a drainage hill.

Forming a Hill

In the wild, succulents are often found growing out of rocky cliffs and slopes that allow rainwater to drain away quickly. When designing your garden, you'll want to mimic nature by building up a hill lined with rocks to help hold the soil and plants in place.

Once I lightened the soil with amendments, I refilled the bed and continued piling the mix against the wall where a hill began to form. I used my hands to shape the slope and level the highest point like a miniature mesa. I then outlined the flattened area with rocks to help visually delineate the top terrace and repeated the process a few inches lower.

As you form your terraces, be mindful of how and where the rain will flow and drain. You don't want any pools to form in your bed, so be sure you fill it with enough soil, and align any low spots with the gaps between your border materials. You can create natural rivulets with stones to help guide water down the slope and towards the spaces you left for it to exit the bed.

Selecting Succulents for Planting In-ground

I first made sure the succulents I wanted to plant in the ground were acclimated to the sun by leaving them in their pots near the bed for about a week (*see 'Chapter 3—Succulent SOS: Take Action - Increase Light'*). Those that showed signs of duress from too much full sun didn't make the cut and were returned to the porch. It was only early spring when I started working on the bed, and by mid-summer they would have been cooked to a crisp.

Planting succulents into the hill at an angle between half-buried rocks rather than straight up and down helps prevent water from standing on leaves and lifts roots well above ground-level. Cold-hardy varieties like Sempervivums and Stonecrop Sedums can withstand freezing temperatures when planted with this method because their roots remain high and dry.

My garden is slightly covered by the roof, but it still gets rained on and sometimes it gets too much rain for comfort. That said, I've only lost a clump of

1. Pile dirt at the back of the bed to make a hill and pat it down to level it out. Line it with rocks and plant your taller specimens towards the back of the bed.

2. Creating slopes helps the rain to run down and out of the garden bed. You can use rocks to form a 'rivulet' for rain to follow down the slope and out through the space you leave between your borders.

3. Borders should allow for water to leave the bed. You can use large rocks, garden bed borders, and bricks among other items.

4. Planting at an angle in the hill rather than straight up and down helps prevent rain from standing on leaves and lifts the roots above ground-level.

To see the final results and updates on the garden, you can watch the videos on my Youtube channel playlist titled 'In-ground succulent gardening.'

bright and delicate Sedum japonicum, and I'm certain it was from the sun coming out after a morning rain a few days in a row. I also noticed my Echeveria peacockii was struggling from too much water so I pulled it up and repotted it. I think if I'd done a better job of removing the organic soil from around its roots before planting, it would have been happier staying in the ground.

The succulent and cacti varieties that did make the cut and are still going strong include: Sedum adolphii *'Firestorm'* and *'Golden Sedum,'* Euphorbia tirucalli *'Sticks on Fire,'* E. trigona, Graptopetalum paraguayense, Graptoveria *'Fred Ives,'* Echeveria minima, E. *'Neon Breakers,'* E. lilacina, E. *'Perle von Nurnberg,'* E. *'Tippy,'* E. pulidonis, Kalanchoe marnieriana; K. fedtschenkoi *'Lavender Scallops,'* Senecio mandraliscae *'Blue Chalk Sticks,'* Pilocereus azureus, Cereus peruvianus, Trichocereus schickendantzii, Mammillaria spinosissima, Parodia buiningii, Echinopsis oxygona, Opuntia *'Old Mexico,'* and O. *'Santa Rita.'*

Designing the Space

There are dozens of great books on xeriscaping with succulents you can refer to for inspiration, so I'm only going to touch on a few of my own design preferences. I like having the taller cacti and Euphorbia *'Sticks on Fire'* at the back of the bed so it doesn't hide the smaller plants. I also like grouping similar varieties closer together in odd numbers while leaving negative space between the groups. Adding colorful varieties towards the center and at each end of the bed entices the eyes to travel across the xeriscaped panorama. Smaller cacti stand out more when the area around them is left empty or accented with a large rock.

Plan for growth: Add taller plants to the back of the bed and smaller or slower growers to the front.

You can use natural objects like Mopani wood or driftwood as visual accents that can double as shade for smaller Sedums and lift trailing plants off the ground. Large chunks of quartz and slate mimic boulders and cliffs. If you have been saving any special rocks or seashells your succulent garden is the perfect place to display them. Keep in mind that rocks get hot in the full sun so plan accordingly.

Protection

In regions where snow and freezing are probable, you'll want to be able to easily cover your garden with frost cloth, so leave some space between the plants and the wall and at the ends of the bed. You can use garden stakes to secure the cloth in place so the wind doesn't blow them away.

I start to worry about the cold when temperatures drop below 45°F (7°C) for more than two days in a row and I really worry if it has rained enough to penetrate the soil around the plants' roots. When Houston got hit with a cold front in early January 2017 that brought temperatures below 25°F (-4°C) for 3 days in a row, I covered my plants and prayed that it wouldn't rain. Luckily most of them made it through, otherwise I probably wouldn't have expanded the bed.

I love my succulent garden and knew there was a chance I'd lose some plants to the elements. I made sure not to experiment with anything special to me and to choose varieties that had better odds of survival in our crazy weather. It's best to opt for succulents that you've had around for a while, or that are acclimated to your climate. My advice is to start out small and expand on what works each year.

Before the expansion
Gardens are always a work in progress...
Opuntia 'Old Mexico.'

Chapter 8
TASKS &
PROJECTS

"Start where you are, with what you have. Make something of it and never be satisfied."

—*George Washington Carver, American agricultural scientist and inventor (1864-1943)*

Potting Up Plants

Whether you're potting up a single succulent or multiples, these tips will help you keep them healthy and pleasing to the eye. You'll need a container with drainage holes, potting mix, and plants *(see 'Chapter 1—Basic Tips: Soil' and 'Containers')*. You may need to cover larger holes with a piece of screen or plastic mesh.

In most cases your pot should be tall enough to allow for a few inches of root growth and just wide enough to contain the plants. If you use a pot that is too large for the size or quantity of plants, the soil will take too long to dry which increases the risk of rot. Too small of a pot will cause your plant to stop growing and may create problems with the root system *(see 'Chapter 1—Basic Tips: Containers - Pot Size')*.

Top dressing is optional but recommended to help give your pot a more finished look and keep the lower leaves of your plants above the soil. You can use pumice, Turface, expanded shale, and other porous rocky materials.

Start by filling your container about halfway with your potting mix. Depending on the length of the plants, you may need to add or remove soil so they'll stand upright. While holding the stem just above the roots, start adding more soil around the plant and pack it down gently. There should only be a half inch (1.3 cm) of stem or less beneath the soil and an inch or less between the lowest leaves and the top of the pot. Leaving too much space allows water to pool and not leaving enough will cause the soil and top dressing to spill out of the pot when watering.

Supplies: Pot with drainage hole, screen, potting mix, top dressing material

Single plants: When potting an individual succulent, it is even more important not to overpot it. You may need to increase the amount of drainage materials to help it stay in place. A layer of top dressing will really help the plant stand out. I like adding a pretty rock, crystal, or other natural accessories like petrified wood to give it a little something extra.

Multiple plants: If you're having trouble getting a plant to stay where you want it, start placing other plants around it for support then add soil and drainage materials until everyone is snug. It's okay to pot succulents close together as most are slow growing, but you'll want to use varieties that have similar care needs. Potting a cactus with a bunch of Echeverias looks great, but their watering needs are different and problems could ensue. Certain cacti like Opuntias are better choices since they don't mind a bit more water than other genera.

Decide which direction will face forward and pot taller plants towards the center or back of the pot so they don't hide the shorter plants. Even if they're all of equal size when you're creating the arrangement, plan for future growth. I like potting rosettes at angles so they're already 'leaning' towards the light. With this method, you can create a mounded look around a central taller plant. I also like to leave negative space between plants by using top dressing to create contrast and give the eyes room to wander.

Potting up multiple plants:

1. Hold the plant above the roots and below the leaves to protect the farina, if present.
2. Tilt the pot while you work to help keep your plants in place.
3. Start adding more plants and soil. You can reposition them as needed.
4. Once your plants are where you like them, add your top dressing and smooth it around
5. Use tweezers, soft brushes and your breath or a bulb aspirator to clean the plants.
6. Finished! Just wait a few days to water to give any broken roots a chance to heal.

Projects

Wondering what to do with all of your minis?

1. Tree of Life mini mosaic
2. Corktus Magnet
3. Nesting Pots
4. Saucer of Love

1. Tree of Life mini mosaic

Supplies
Small pot filled with potting mix
Top dressing
Tweezers
LOTS of mini propagations

This is a lovely project that I find meditative. Begin by thinking of a design you'd like to make and start outlining it with your tiny plants. Fill the outline with more plants using the tweezers to avoid knocking the others out of place. Once you're finished, fill in the negative space with your top dressing of choice. Enjoy watching the wee succulents grow until you're ready to move them to a bigger pot.

2. Corktus Magnet

Supplies
A cork
Magnet
Drill, corkscrew, or screwdriver
Glue gun
A little bit of potting mix
Damp Sphagnum moss
Tweezers
One or more tiny succulents, preferably rooted with a stem

First you'll need to hollow out the cork. I used my drill and it made the job super easy. I also added a smaller hole at the base for drainage but with such a small 'pot' there's no need to worry about overwatering. You can use a corkscrew or screwdriver as well. Attach the magnet with the glue gun and let it dry for several hours or more. Fill the hole about halfway with soil and add your plant. Use the tweezers to pack the damp moss around the plant. If you make little balls of the moss, they will expand once in the hole. Water weekly or so by holding the cork under water for a few seconds.

3. Nesting Pots

Supplies
Pots of various sizes
Potting mix
Top dressing
Tweezers
Tiny succulents, rooted or unrooted

Begin by filling the largest pot about 3/4 full with potting mix. Fill the next pot the same way and place it in the first pot. I like the offset look of it closer to the rim but you can center it if you'd like. Go back and fill the largest pot with your mix, snugly burying the base of the second pot. Repeat until you have your stacked pots in place then begin adding plants. Finish with your favorite flavor of top dressing.

4. Saucer of Love

Supplies
A clay saucer
Potting mix
Pretty rocks, crystals, etc
Succulents, rooted or unrooted

This is super simple but loads of fun to make! Fill the saucer 3/4 full of potting mix. Add your decorations then fill in the negative space with your pretty little succulents. To water, use a spray bottle to soak the soil and keep the plants in place (about once a week.)

Hand-painted mini pots for mini sucs

5

7

Let's hang out!
5. Burlap Sack of Sucs
6. Insta-hanging pot
7. Flying Saucer
8. Kokedama Ball

6

8

5. Burlap Sack of Sucs

Supplies
A burlap sack (mine were from coffee)
Garden twine
Sphagnum moss and/or coco coir
Tweezers
Succulents with a stem

First you're going to fill the sack with the moss and/or coco coir until it's packed almost to the top. Choose plants with a longish stem if possible. Lay the sack flat and put your non-dominant hand inside, palm facing up. Use the tweezers to spread the burlap so you have a hole large enough to stick the stem through and use your hand that's in the sack to gently pull it in further. Repeat until you have all of your plants in place then secure the top with more twine. You can carefully test the sack to see if the plants stay in place or wait a week or more if you'd like to be really sure. To water, lay it flat and use the shower setting on a hose (once every other week should be fine). *Optional*: To make the sack sit upright, before adding the moss, fill the bottom with your choice of drainage materials to weigh the bottom down.

6. Insta-hanging Pot

Supplies
A pot
Drill with a ceramic bit
A towel
Water
3 pieces of ribbon or garden twine

Mark 3 holes around the rim of the pot. Stuff the pot with the towel and lay it on its side. Dampen the rim then drill each hole. Thread the ribbons through each hole from the outside in, then secure it with a knot at the end large enough to not fit through the hole. Finally, tie the three ribbons together at the top. It's up to you how long you want the hanger.

7. Flying Saucer

Supplies
A clay saucer
Sphagnum moss
String or garden twine
Succulents, preferably rooted with a stem

Take a big piece of damp moss and press it into the saucer. Place the saucer on top of a long piece of twine and tie it firmly to bind the moss. Turn the saucer and repeat until the moss is contained. Make a knotted loop through one of the sections somewhere near the rim for hanging the saucer later, then tuck the succulents into the moss. You'll need to wait at least a week to check to see if they've rooted well enough to hang the saucer. When ready, add a piece of twine or ribbon to the loop you made and hang it somewhere in bright indirect light. To water, lay it flat again and gently pour water over the moss. Let it soak until wet all the way through then hang it back up.

OOPS! I like to save my broken pots to make sections in larger pots.

8. Kokedama Balls

Supplies
Garden twine
Sphagnum moss and or coco coir
Garlic sack or other plastic netting
Potting soil
Succulents, rooted or unrooted with a stem

This is a bit tricky but I've found a way to make it easier by using the plastic netting from garlic, small potatoes, etc. Fill the netting with potting soil until you have enough to form a ball—the size is up to you but 3-5" in diameter is ideal. Trim the excess netting but leave a few inches above the soil.

Start placing your succulents in the opening of the sack with the stems in the soil. Use the twine to secure them in place then fold the netting down or trim it away.

Next, take a big piece of damp moss and flatten it out until you have enough to cover the ball. You can press several pieces together. Set the sack in the center of the moss and fold the moss up and around the sack and stems.

The last step is to secure the moss in place with the twine. Begin by wrapping the stems as low as possible with twine and tie them off. Tie your loose end of twine to the same place and tie it off, but keep it attached to the bale.

At this point you'll start wrapping the ball with one hand while turning it with the other. Start with a diagonal over-under motion like you're winding up a yo-yo, but with each time around, turn the ball about an inch. It's the same movement used to wind up a ball of yarn. Continue until it's wrapped all the way around and don't worry about covering the whole thing perfectly since seeing the moss peeking through adds texture and a natural look.

Tie it off and add another piece of twine to hang it from. To water, soak the whole ball (while keeping the plants above water) 1-2 times a month and hang to dry. These can last a year or more before they'll need to be unwound and repotted.

*Echeveria derenbergii,
Sedeveria 'Starburst,'
String of Raindrops*

Ledebouria socialis

Gifting and Shipping

Above: A gift basket for a succulent lover includes pots, a nailbrush, cuttings, labels & other fun garden knick-knacks.

Left: Getting ready to ship some plant-mail! I usually unpot everything, remove the soil from around the roots, and wrap bundles of plants in packing paper. Sometimes I fill empty plastic pots with plants for extra protection before wrapping them. Fill the bottom and top of the box with more paper or biodegradable packing peanuts.

Cleaning Your Plants

You may need to clean your succulents after repotting them or as dust, dirt, and sometimes cobwebs accumulate over time. Farinose varieties are harder to clean without smudging the farina but since their leaves repel water, a squirt bottle or gentle hose setting is usually the best cleaning tool for these types. Be sure to keep them out of direct sunlight until they dry.

Other succulents can be handled and wiped without worry. You can use water, a damp cloth, soft brushes, and tweezers to spot clean.

I really like my little bulb aspirator and use it to blow the dirt and dust off of delicate leaves and from between cacti tubercles and spines. It came in a succulent toolkit and I use it quite often.

Drilling for Drainage

Since all pots should have drainage holes, you should have a power drill! With the right bits, almost anything can be drilled. Look for carbide or diamond-coated bits and pick up a few since they dull after several uses. Hole saw bits are great for making larger drainage holes.

In addition to the power drill and correct bit, you'll need a towel or grassy spot to work over, water and a container for drilling.

First determine where the holes should be placed. You can mark the bottom of the pot with a pencil or a gentle nick from the bit. Invert the container and dampen the bottom with water then align the tip of the bit with the spot you want to drill. Start on a lower speed to make sure you don't crack the container and only use the weight of the drill itself to add pressure. Once you get a little dimple drilled, you can turn up the power level and apply more pressure. Be sure to keep the bottom damp to help cool the bit and contain the dust. You'll be able to hear when the bit is almost through the hole. At that point, ease up on the pressure or the drill might hit the pot.

Repeat until you have several holes. Now just add a piece of screen to help keep the soil in the pot if you'd like and enjoy your handiwork!

Note that drilling metal from the bottom will create raised divots inside the container which may prevent water from draining. You can use a hammer or something heavy to tap them flat but be careful since they'll be sharp.

DIY Shelves

If you're in need of more shelf-space (aren't we all?) it's fairly simple to create your own at an affordable cost. I use cinder blocks and pressure-treated boards painted with outdoor deck stain.

The blocks are less than two dollars each and a 2" x 6" x 16' board is around ten dollars. Most shops will cut the boards for you at no additional charge. While this set-up is very heavy, it is cheap, effective, and weather-proof.

Use two blocks for every 5 feet of board and opt for wider boards so they can hold larger pots.

You can also use inverted boxes like these stacked wine crates I found for free. The options are endless and with a bit of imagination, you'll be seeing potential shelves everywhere!

DIY Plant Stands

Small chairs
Stools
Side tables
Upside down pots
Wood blocks
Cinder blocks
Stacked bricks
Stacked books
Large jars

Logging and Journaling

I highly recommend keeping a log of your collection. It can be as simple as the date you bought a plant and its name, or it can include more details like care needs, where it originates, where you bought it, and where you identified it.

Journaling is another great way to keep track of all sorts of garden activities such as when you planted something or started a set of propagations. Being able to look back helps track progress, flowering periods, and growth cycles, as well as seasonal concerns like pests and flooding. It's like writing your own personal gardener's almanac.

I use a spreadsheet with different tabbed pages dedicated to my succulent log, non-succulents, and my journal. See 'Chapter 5—Identification' for an example. While a spreadsheet allows you to sort the columns and quickly search for an entry, notebooks are great too. You can use the Plant Journal at the back of this book to start keeping track of your collection until you run out of space. By space of course I mean pages, not room for plants.

Another way I log my plants is by photographing them and saving the pictures to an online service. Photograph the side and top views of your plants along with any flowers or other notable characteristics. While some phones are capable of producing high quality shots, I use a digital camera for my photography and upload the photos from an SD card using the desktop version of Flickr. I recommend you create albums based on your plant types and title the pictures as you upload them so they're easy to search for down the road. Retake photos of the same plants every few months and in time you'll have a lovely and informative gallery of your collection to reflect back on.

Leaving Town

The day may come when you need to say goodbye to your succulents for a while. If you're going to be gone for 10 days or less, preparing your collection for your departure may be as simple as giving them a drink before you go. If you'll be gone longer, you're probably planning on having a plant-sitter stop by to care for them. for them instead.

-Decide on the days you want the plant-sitter to water your succulents before you leave. The soil should dry between waterings so every 7-10 days is a good schedule.

-Show them how you water or make them a video. If they have to water during the daytime, ask them to be cautious of wetting the leaves.

-If you have any plants that don't need to be watered while you're gone, write NO on a slip of paper and use a rubber band to secure it to their pots.

-Ask them to inspect your plants for pests and decide on a location where they can quarantine any that look suspicious. If they're comfortable with treating them for pests, give them a quick lesson and show them where to find the necessary supplies.

-Invite them to send you a photo of any plant they have questions about.

A local sale in my backyard

Selling Succulents

The time may come when you have so many succulents you'll need to find them new homes. I've sold my excess plants locally and online, and you can too! Some states have different laws about the types of plants being shipped and whether or not you would need a license to sell them. There are also plants that are patented and propagation is prohibited, so do a bit of research to be on the safe side.

As far as pricing goes, there's no reason to think you should offer your plants at the same cost as a professional seller or big box store. Think about how much you would pay for what you're selling and assume buyers will too. You'll need a way to accept credit card payments so be sure you have a Paypal account or a similar alternative.

Be diligent about thoroughly inspecting each plant you plan to sell for pests or problems from top to bottom—roots and all.

Local Sales

Selling locally is great because you don't have to worry about shipping them. I've had a couple of plant sales in my backyard and sold some during a garage sale. Being able to make some extra cash from something I had grown really made me feel proud of myself.

Whether you're selling cuttings or potted plants, you'll need some containers to help your customers get their purchases home safely. Start saving empty boxes and paper to stuff around pots to keep them from shifting during transport.

If you hold an in-person sale, you'll need to label the pots with prices in a way that can easily be removed. Don't use tape on terracotta pots as it can leave a sticky residue behind. I liked using a piece of paper affixed to the pots with a rubber band.

Your customers may have questions or just want to chat about succulents so it's a good idea to have someone help you handle the sales. Consider printing out some copies of general care tips and maybe even direct them to this book—*wink wink!*

If you don't want to go through the trouble of setting up a yard sale, you can post your offerings online in a number of places like Craigslist, Facebook Marketplace, and on your personal social media accounts. Once someone confirms their interest, they should be ready to pay and reserve their items. Paypal has an invoicing system that will email the buyer with a link to enter their payment information.

You can arrange a place to meet the buyer for delivery, or if you're comfortable with them coming to your home to pick up their purchase, have the items neatly tucked in a box, and consider taking a photo of the plants for your records. This helps you remember who bought what and documents the condition the plants were in at the time of the sale.

Online Sales

While it takes some extra effort to package your plants, the good thing about succulents is they ship really well if you wrap them with care. I always use the insured Priority cubic rate option through USPS and haven't had any problems with them arriving damaged. Be sure the customer double checks their shipping information and triple check what you write on the label. Send them the tracking number and keep a record of your sales. Be sure to check the forecast of the zip codes you ship to, and avoid mailing anything during times of the year when it's too hot or too cold.

You'll need a good amount of paper or packing peanuts to line the box so nothing shifts around. Of course it's optimal to use recycled materials so ask around to see if friends or family have anything that works. The bottom, top, and inner-sides of the box should have a nice layer of padding. I use wadded up packing paper and peanuts if I have any laying around. Avoid using plastic bags since they can trap moisture and don't allow the plants to breath.

I always ship bare-rooted plants because they're easier to pack and most succulents can tolerate a couple of weeks without soil or water. Remove as much dirt from the roots as possible so everything arrives neat and tidy.

Lay out a piece of paper and place the plants in the center. Wrap them loosely then wrap them again with another piece of paper before securing the edges with tape. You can even add more peanuts or crumpled paper around the plants before wrapping them up for extra protection. Place your bundles in the box then add as much padding as you can to the sides and top. For more fragile succulents, you may need to put them in a smaller box or empty plastic pot lined with padding.

Chapter 9
BUYING GUIDE

"A garden is a grand teacher. It teaches patience and careful watchfulness; it teaches industry and thrift; above all it teaches entire trust."

—Gertrude Jekyll, British horticulturist, 1843-1932

Where to find succulents for sale

Locally

Of course big box stores sell succulents, but the varieties they stock doesn't change much so eventually you'll need to look elsewhere for more unique options. And while some of these chain stores take better care of their plants than others, it is very common to see pest-ridden and overwatered succulents already on their deathbeds being sold at full price.

Another thing to watch out for are plants that have been altered with paint and hot-glued flowers. Yes, that's right. Those neon pink Haworthias are coated in a tacky photosynthesis-blocking paint that doesn't wash off. And those flowers on cacti could very well be dried and dyed strawflowers affixed to the top of the living plant with glue. If in doubt, touch it. You know what a real flower feels like. Most cacti flowers will wilt and fall off before getting a chance to dry so it will be very obvious to the touch. Removing the glued on flower usually takes part of the plant with it. You will rarely find these sad and defaced succulents at any nursery who wants to be taken seriously, and as more people become aware of why it's bad for the plants and stop buying them, we should start seeing them replaced with more desirable inventory.

As someone who lives in a large city with a climate not terribly kind to succulents, the local options are fairly slim. But I keep looking and the list is growing longer as sucs grow more popular. Besides searching the internet, you can call around and ask local nurseries if they have succulents or can recommend someone who does. There are also tons of regional-based succulent forums and groups on Facebook and elsewhere. Join and ask where members near you have purchased quality plants.

Also keep an eye out wherever you see plants for sale. I found a really great selection in an Asian grocery store, along with some veggies and pepper plants. Another place with the vague name of JRN turned out to be a Vietnamese-owned plant oasis, with everything from bonsai trees and Monstera adansonii to weird caudex plants and some of the biggest Echeveria I'd seen. Needless to say I was in heaven and I spent all of my money very willingly.

If you're lucky, your city may have a local cactus and succulent society. If so, they usually have seasonal sales and that is an awesome chance to pick up some interesting plants that are already accustomed to your climate.

Buying Online

Some shops are quicker to ship and offer better customer service than smaller or individually owned stores, but the latter tend to have a more diverse selection. You can find reputable sellers on Instagram, Facebook groups, succulent forums, as well as ask about others' opinions on a shop you are considering purchasing from.

Imported Succulents

There's a reason imported succulents are so expensive. First, they may be rare to begin with, so factor in the survival rate of being shipped overseas with the chance of Customs seizing the entire package, and you'll understand why some come with such a high price tag. Always request a phytosanitary certificate or you are running a huge risk in your order being confiscated and destroyed. If you don't want to take the full risk of losing your money, you should buy from an importer located in your country. You'll be paying more but it will be a sure thing.

Beginner Friendly Succulents and Others

There are plenty of easy growers I like to recommend to people new to keeping succulents, and many have been included throughout the book because they are still some of my favorite plants. Don't be afraid to try more advanced varieties once you get the hang of some of those from the first two lists. You'll notice there isn't a list for full sun since most succulents will not thrive in direct light all day unless you're in a mild climate.

Rosettes

Aeonium *'Kiwi'*- *p.57*
Anacampseros rufescens- *p.54*
A. telephiastrum- *p.139*
Echeveria *'Lola'*- *pp.17, 55, 96-97*
E. *'Neon Breakers'*
E. *'Perle von Nurnberg'*- *pp.39, 138*
E. *'Tippy'*- *p.14*
E. *'Topsy Turvy'*- *p.97*
E. lilacina- *pp.123, 138*
E. minima
E. parva- *pp.41, 122-123*
E. pulidonis- *pp.63, 88*
E. pulvinata- *p.138*
E. subsessilis
Graptopetalum paraguayense- *pp.16, 38, 127*
Graptosedum *'California Sunset'*- *p.145*
Graptoveria *'Debbi'*- *p.84*
G. *'Fred Ives'*- *pp.30, 145*
Sedeveria *'Starburst'*- *pp.17, 84, 132*
Sedum adolphii *'Firestorm'*- *p.130*
S. adolphii *'Golden Sedum'*
Sempervivum species- *p.78*

Others

Agave lophantha
Aloe barbadensis (Aloe Vera)
A. juvenna
A. maculata- *p. 92*
Crassula ovata varieties- *p.110*
C. peruvianus- *pp.45, 88*

Echeveria pulvinata, E. lilacina

Echeveria *'Perle von Nurnberg'*

More Advanced

These varieties are more particular about their care requirements or need a more controlled environment to remain healthy.

Anacampseros telephiastrum

Fenestraria rhopalophylla
'Baby Toes'

Orostachys boehmeri *'Chinese Dunce Caps'*
Pachyphytum oviferum *'Moonstones'*
Sedum morganianum *'Burro's Tail'*
S. rubrotinctum *'Jelly Beans'*
Sempervivum *'Oddity'*- *p.112*
Senecio haworthii *'Woolly Senecio'*
S. rowleyanus *'String of Pearls'*
Titanopsis calcarea- *p.140*

Part Sun

Aloe aculeata- *p.144*
A. aristata
A. barbadensis (Aloe Vera)
A. distans
Beaucarnea recurvata *'Ponytail Palm'*- *p.143*
Disocactus ramulosa
Dorstenia foetida- *p.144*
Echeveria pulvinata- *p.138*
Euphorbia bupleurifolia x suzannae
E. decaryi
E. ingens
E. polygona *'Snowflake'*- *p.50*
E. trigona- *p.99*
Haworthia attenuata *'Zebra Plant'*
H. limifolia
H. coarctata- *p.101*
Huernia zebrina *'Lifesaver Plant'*
Hylocereus undatus *'Dragon Fruit'*
Kalanchoe beharensis *'Fang'*- *p.140*
Kalanchoe beharensis minima- *p.111*
K. millotii- *p.111*
K. orgyalis *'Copper Spoons'*
K. rhombopilosa- *p.67*
Ledebouria socialis *'Silver Squill'*- *p.128*
Opuntia monacantha monstrose variegata
Plectranthus tomentosa
Portulacaria afra *'Elephant Bush'*- *p.140*
Rhipsalis species
Sansevieria cylindrica
Senecio radicans *'String of Bananas'*
S. radicans *'String of Raindrops'*- *pp. 72, 113*
Stapelia grandiflora

Titanopsis calcarea

Kalanchoe beharensis *'Fang'*

Portulacaria afra *'Elephant Bush'*

Cold-hardy

This list is short because there aren't many succulents that can survive a freeze, but several varieties are able to make it through a frost or freezing temperatures if kept dry.

Agave havardiana- *p.78*
Agave ovatifolia
Aloe brevifolia- *p.93*
A. ciliaris
A. ferox
A. humilis- *p.141*
A. juvenna
A. striatula
Ariocarpus fissuratus
Astrophytum myriostigma
Dasylirion texanum
Delosperma species
Echinocactus grusonii
Echinocereus dasyacanthus
E. rigidissimus- *p. 141*
Euphorbia resinifera
Ferocactus wislizeni
Fouquieria splendens
Hesperaloe parviflora
Mammillaria heyderi
Manfreda virginica
Opuntia *'Old Mexico'*- *p.119*
Oreocereus trollii
Sempervivum species- *pp.78, 112*
Stonecrop Sedum species- *p.112*
Yucca aloifolia
Y. rupicola

Hot and Humid

Aloe maculata- *p.92*
Anacampseros rufescens- *p.54*
A. telephiastrum- *p.139*
E. *'Topsy Turvy'*- *p.97*
E. lilacina- *pp.123, 138*
Echinopsis schickendantzii- *p.144*
Epiphyllum anguliger *'Ric Rac'*- *p.82*
Euphorbia suzannae- *p.99*
E. tirucalli *'Sticks on Fire'*- *p.145*
Graptoveria *'Debbi'*- *p.84*
G. *'Fred Ives'*- *pp.30, 145*

Aloe humilis

Echinocereus rigidissimus

Haworthia limifolia
Kalanchoe beharensis *'Fang'- p.140*
Kalanchoe rhombopilosa
K. synsepala
Ledebouria socialis *'Silver Squill'- p.128*
Mammillaria spinosissima- *p.115*
Opuntia ficus-indica
O. *'Old Mexico'- p.119*
Parodia werneri- *p.142*
Portulaca molokiniensis- *p.145*
Sansevieria cylindrica
Sedeveria *'Starburst'- pp.17, 84, 132*
Senecio crassissimus *'Lavender Steps'*
S. radicans *'String of Bananas'*
Stapelia gigantea
S. grandiflora
Most Kalanchoes
Most Haworthia
Most Euphorbia
Epiphytic jungle cacti

Non-succulents you'll also love

I joke that one of the best ways to prevent overwatering is to include some thirstier plants in your collection. Some of these are still slow growers, but others will add a bit more activity to your shelves.

Begonias
Bromeliads
Carnivorous plants
Coleus
Dracaena
Monstera adanosii *'Swiss Cheese Plant'*
M. deliciosa
Philodendron
Pilea peperomioides- *p.81*
Pothos
Tillandsia *'Air Plants'- p.75*

More from my collection on pages 144-145

Parodia werneri

Kalanchoe gastonis-bonnieri
'Donkey Ears'

Kumara plicatilis
'Fan Aloe'

Supplies

These are some of the supplies you may need over time.

Soil & Drainage Materials

Topsoil
Perlite
Pumice
Turface/Porous ceramic soil conditioner
Expanded shale
Lava rock/scoria

Potting

Unglazed clay pots with drainage holes
Clay saucers
Screen to cover larger drainage holes
Small plastic nursery pots

Cotyledon tomentosa
'Bear Paws'

Tools

Pitcher with a small spout
Big bins for holding soil/drainage materials
A shallow bin to work over for easy clean up
Scoops of various sizes
Tweezers
Sifter for making seedling soil
Trays for propagations
Broom and dustpan, full size and hand-held
Brushes & bulb aspirator for cleaning leaves

Pests

Rubbing alcohol
Neem oil
Other pesticides if necessary
Spray bottle
Bins for treating and quarantining

Grow Lights

Hanger or stand
Power supply
Timer
Table
Trays for catching water
Fan

Beaucarnea recurvata
'Ponytail Palm'

Albuca spiralis

Aloe aculeata

Cereus forbesii *'Spiralis'*

Crassula *'Morgan's Beauty'*

Dinteranthus sp.

Dorstenia foetida

Echeveria *'Chroma'*

Echeveria *'Ruffles'*

Echinopsis schickendantzi

Euphorbia tirucalli

Graptopetalum bellum

Graptosedum *'Cali. Sunset'*

Graptoveria *'Fred Ives'*

Jatropha berliandieri

Kalanchoe tubiflora

Kalanchoe laetivirens

Portulaca molokiniensis

Senecio pendens

KNOWLEDGE BANK

"I know of no pleasure deeper than that which comes from contemplating the natural world and trying to understand it."

—David Attenborough, English broadcaster, natural historian

Succulent Anatomy

The following information will help you identify your plants and understand what observations to make to determine various traits. You'll learn about the anatomy of succulents and cacti and the terms used to describe their characteristics.

For practice, I suggest you use one your favorite succulents as a subject to study while reading through the Knowledge Bank. Be sure to refer to the glossary section for definitions. While this is not a complete botanical guide, it provides a great starting point to build on.

Pleiospilos nelii

Leaves
Morphology (Shape) and Phyllotaxis (Arrangement)

Shape: The general form a leaf exhibits

acicular acuminate cordate cuneate elliptic falcate hastate lanceolate

linear obovate obtuse orbicular ovate rhomboid spatulate

Arrangement: How leaves are attached to the stem

simple compound alternate opposite rosette palmate whorled

Margins: The main characteristic of the outer edge

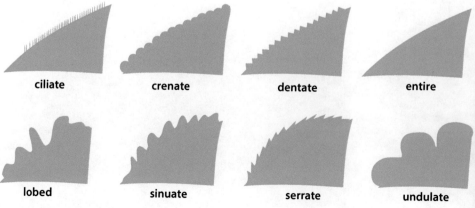

ciliate crenate dentate entire

lobed sinuate serrate undulate

Shapes

acicular: Needle-like
acuminate: Tapering gradually to a point
cordate: Heart-shaped
cuneate: Wedge-shaped
elliptic: Oval-shaped
falcate: Hooked; curved like a sickle
hastate: Narrowly triangular; like a spearhead
lanceolate: Shaped like a lance head; oval shaped, tapered to a point at either end
linear: Long and narrow like a grass blade
obovate: Egg-shaped with the narrow end attached to the stem
obtuse: Having a blunt tip
orbicular: Circle-shaped; rounded
ovate: Egg-shaped with the wide end attached to the stem
rhomboid: Diamond-shaped
spatulate: Shaped like a spoon

Arrangements

simple: Having one leaf blade
compound: Having two or more leaves
alternate: Leaves growing one per node and alternating sides
opposite: Leaves growing in pairs at the same level attached to the same node
rosette: Overlapping arrangement of leaves in a spiral
palmate: 3 or more leaves radiating from a single point
whorled: 3 or more leaves radiating from a single node

Margins

ciliate: Having small hairs
crenate: Featuring rounded teeth; scalloped
dentate: Toothed
entire: A simple leaf margin that is smooth with no lobes, notches, or ridges
lobed: A partially rounded portion of a leaf margin
sinuate: Margin with wavy indentations
serrate: Saw-toothed with teeth pointing towards leaf tip
undulate: Shallow waves with wider spacing than sinuate margins

Terms for describing surfaces of leaves and stems

Glaucous, farinose, punctate, tuberculate, glabrous, pubescent (among others)

Some types of pubescence (surface hair)

arachnoid- Cobweb-like; soft entangled
bearded- Long, stiff
bristly- Stiff
ciliate- Fringe along a leaf margin
floccose- Soft, woolly tufts
hirsute- Rigid, rough to the touch
hispid- Rigid, may pierce skin
hoary- Covering of short, fine hairs
pilose- Soft hairs
scabrous- Short, stiff hairs
tomentose- Dense matting of soft white hairs
villous- Shaggy, long, soft, not matted
wooly- Long, soft entangled hairs

Observations in succulents

Stem characteristics

Erect, decumbent, branching, width, length, visible, hidden, caudex, leaved or leafless, herbaceous, woody

Leaves

Arrangement- Alternate, opposite, whorled
Traits- Size, form, margin, upper/lower surface, quantity, venation (veins)

Inflorescence (The entire flower)

Types- Spike, raceme, compound raceme, umbel, compound umbel, cyme, cinncinus, thyrse (among others)

Traits- Direction of growth, length and width of stem, color, leaved, bracts, branched or unbranched, pedicels

Sex- Male, female, or perfect, meaning the flower contains both male and female reproduction organs

SEMPERVIVUM
WULFENII

S. EDWARDS

1–5. Entire inflorescence
1. Corolla/flower
2. Peduncle/flower stalk
3. Petals
4. Stamen

5. Anthers
6. Offsets/chicks
7. Stolon/runner
8. Scale leaves

Parts of a Sempervivum wulfenni Illustration by Sydenham Edwards, Welsh botanical artist (1768-1819)

Observations in flowers

Petals- Inner and outer color, shape, size, quantity
Stigma- Color, visibility, presence
Pistil- Presence, color, length
Stamen- Presence, color, quantity, length, position
Anthers- Color, quantity, shape
Sepals, tepals- Presence, color, shape, quantity
Fragrance- Presence, sweet, citrus, carrion
Pollen- Presence, color

Pleasantly fragrant flowers

Aloe conifera: citrusy, grapefruit
Epiphyllum oxypetalum: heady perfume
Euphorbia obesa: citrusy, lime
Hoya carnosa: sweet
Lithops salicola: honey
Mammillaria plumosa: sweet, fruity
Selenicereus grandiflorus: orange blossom
String of Bananas/Pearls: light cinnamon

*Left: Parts of an Opuntia ficus indica
Illustration by Otto Wilhelm Thomé,
German botanical artist (1840-1925)*

A. Cladode/stem/pad/segment
1. Flower
2. Stamens
3. Style
4. Stigma
5. Fruit
6-7. Seed

*Right: Various cacti from
Curtis's Botanical Magazine*

Observations in cacti

Stem habit (form, shape of main body)- Angled, button, columnar, cylindrical, flattened-leaflike, flattened-padded, globose, ovoid, prostrate, shrubby, shrubby-treelike, sprawling, treelike, tuberculate

Other traits- Tubercles, areoles, spines, glochids, flowers, fruit, seeds, roots

Various cacti stem habits
(Appearance, shape or growth form)

1. Globose, clumping; Neomammillaria palmeri
2. Angled; Acanthocereus tetragonus
3. Flattened, padded; Opuntia ficus-indica
4. Cylindrical; Cylindropuntia imbricata
5. Button-like; Echinopsis cinnabarina
6. Tuberculate; Coryphantha bumamma
7. Globose, pyramidal; Echinocactus ellipticus
8. Globose; Thelocactus hexaedrophorus
9. Columnar; Coryphantha clava
10. Sprawling; Lepismium lumbricoides

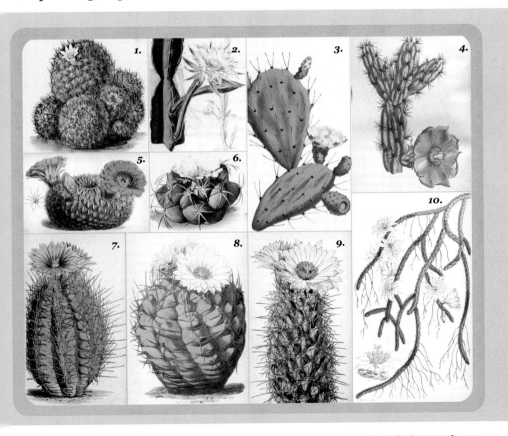

Glossary

-A-
acicular: Needle-like
aculeate: Armed with prickles, e.g. the leaves of Aloe aculeata
acuminate: Tapering gradually to a point
aerial roots: Roots that grow above ground; common in humid climates
Aizoaceae: A plant family of angiosperms including Lithops, Titanopsis, and Delosperma
alternate: Leaves growing one per node and alternating sides
angiosperm: A flowering plant; a plant with seeds enclosed in an ovary
anther: Pollen-bearing part of the stamen
areole: Where spines, glochids, and flowers emerge in cacti
aristate: Bristled tip, e.g. the leaves of Aloe aristata
Asphodelaceae: A plant family of angiosperms including Aloe and Haworthia
Asteraceae: A plant family of angiosperms including Senecio and Othonna
attenuate: Gradually narrowing
axil: The angle between the upper side of a leaf and the stem from which it grows
axillary: Emerging from a leaf axil
axis: The main stem of a plant or inflorescence

-B-
bare-rooted: A plant that has been removed from soil and its roots cleaned of soil
barbed: Having rear-facing points
barbellate: Having barbed hairs
bilateral: Arranged on opposite sides as in leaves on a stem
binomial: A two-part scientific name formed from the genus and species names
bloom: The white coating on a farinose, pruinose, or glaucous plant; a flower
bract: A modified leaf at the base of a flower
bristle: Straight, stiff hair

-C-
Cactaceae: A plant family of angiosperms including cacti
cactus: Plural 'cacti'; plants that have succulent stems and branches, typically with scales or spines instead of leaves
callus: Hardened protective tissue that forms on or around a wound
calyx: A collective term for the sepals of one flower; the outer whorl of a flower, usually green
caudex: Plural 'caudices'; the stem or basal stem of a plant
caudiciform: Having a swollen, thick base in proportion to the whole plant, e.g. the base of Beaucarnea recurvata 'Pony Tail Palm'

central spines: The spine of a cacti radiating from the center of an areole; can be one or more or absent

cilia/ciliate: Having small hairs, e.g. the leaf margins of many Sempervivums and Aeoniums are ciliate

clone: An identical plant derived from a parent plant through vegetative reproduction or propagation, e.g. the new plants formed by leaf propagation are clones

columnar: Shaped like a column as with some cacti

compound: Having two or more leaves

cotyledon: Primary leaves of a plant embryo; the seed-leaf

cordate: Heart-shaped; used to describe the shape of a base of a leaf with the concave portion attached to the stem

corolla: The petals of a flower collectively

Crassulaceae: A plant family of angiosperms including Crassula, Echeveria, and Kalanchoes

Crassulacean acid metabolism (CAM): A photosynthesis process used by plants in order to adapt to arid climates

crenate: Featuring rounded teeth; scalloped

crenulate: Finely scalloped

cross: Hybrid or hybridization; interbreeding between two different plant genera or species

cuneate: Wedge-shaped, describing a leaf

cuticle: Waxy protective exterior layer

cutting: A portion of a plant removed for propagation

cv.: Abbreviation for 'cultivated variety' or 'cultivar'

-D-

deciduous: Seasonal falling/shedding of leaves, bark, or petals

defoliation: Induced or premature loss of leaves; natural shedding of leaves

dentate: Toothed, as in a leaf margin

denticulate: Finely dentate

dormancy: Natural or induced period of rest in a plant; seasonal inactivity

drainage: The process of excess water moving and exiting the area around a plant

-E-

elliptic: Oval-shaped

entire: A simple leaf margin that is smooth with no lobes, notches, or ridges

epiphyte, epiphytic: A plant that obtains water and nutrients from the air and rain; usually grows on another plant but is not a parasite

etiolation: The act of a plant stretching and leaning towards a light source in the absence of adequate light; can also be caused by disease

Euphorbiaceae: A plant family of angiosperms including Euphorbia

extrafloral: Outside of a flower, such as with extrafloral nectary glands on stems

-F-

falcate: Hooked; curved like a sickle
family: Taxonomic ranking between order and genus
farina/farinose/farinaceous: Mealy textured powder or bloom on a plant
fenestrate: Translucence in a plant, particularly leaves, e.g. the leaves of H. cooperi
fertilizer: Plant nutrients, natural and synthetic
filament: Very fine thread; the stalk of a flower stamen
flora: The plants growing in a particular region
floret: A small flower
flower: The reproductive fruit bearing part of an angiosperm; to bear flowers

-G-

germination: In seeds, the process of developing from a dormant seed into a plant
glabrous: Smooth and hairless
glaucous: Dull gray-green to blue color; covered with a white or bluish waxy or powdery bloom
globose: Spherical; often used to describe the shape of cacti
glochid: Small irritating hairs found in the areoles of many cacti
graft: To artificially fuse one rootless plant to one that is rooted in order to enhance growth, as with a grafted *'Moon Cactus'*

-H-

habit: The characteristic appearance of a plant
habitat: Location where a plant lives or originates
harden off, hard grown: The acclimation of a plant to tougher growing conditions as with seedlings or plants that have been in a more temperate, controlled setting
hastate: Narrowly triangular; like a spearhead
herbaceous: Not woody; green and soft, easily broken
honeydew: The sweet secretions of pests such as mealybugs, scale, and aphids
hybrid: A variety of plant created from the crossbreeding of two different types of plants

-I-

imbricate: Overlapping leaves, petals, bracts, sepals, or other parts of a plant
indirect light: Light that is blocked from directly hitting an object
inflorescence: The whole of the flower including the stalk; flowering
indigenous: Originating from or native to an area
internode: The section of a stem between two nodes

-J-

joint: A segment or section of a stem; a node; a point where a leaf or shoot grows from a stem
jugate: Having parts arranged in pairs, as in pinnate leaves
juicy: Succulence of a plant

-K-

keel: In leaves, a prominent ridge down the center of the blade, like a boat
Kelvin: The unit of measurement for light temperature (color/hue)

-L-

lacerate: Jagged
laciante: Irregularly and deeply divided leaf lobes and margins
lamina: The blade of a leaf or uppermost part of bracts, sepals, and petals
lanceolate: Of leaf shapes, narrow, ovate, lower portion broad while tapering at the tip
latex: The milky toxic fluid exuded by a damaged Euphorbia and other plants
leaf: The flat blade growing from a stem used for photosynthesis
linear: Long and narrow like a grass blade
lobed: A partially rounded portion of a leaf margin
lumen: The unit of measurement of total visible light

-M-

maculate: Spotted
margin: The edge of a leaf blade
mealybugs: Small honeydew secreting pests that feed on succulent leaves and roots
midrib: Central vein of a leaf; midvein
monocarpic: An angiosperm that dies after flowering one time - see 'polycarpic'
monotypic: Having only one type, as in a genus having only one species
morphology: The shape or form of an organism, as in the shapes of leaves
mutation: Alteration in the DNA of a cell that cause permanent character changes in plants such as cresting and color variegation

-N-

native: Originating from but not confined to a region
nectar: The sweet fluid produced by flowers and nectary glands
nectary glands: A gland that produces nectar, common to many Ferocactus
node: Joint; a point where a leaf or shoot grows from a stem

-O-

obcordate: In a leaf, broad and notched at the tip; heart-shaped but attached at the convex end

obovate: Egg-shaped with the narrow end attached to the stem

obtuse: Having a blunt tip

offset: A new young plant that forms on the main parent stem; new growth of a stem or branch

opportunistic grower: A plant that continues to actively grow if the conditions are right

opposite: The arrangement of leaves growing in pairs at the same level along a plant's stem

orbicular: Circle-shaped; rounded

ovate: Egg-shaped with the wide end attached to the stem

-P-

pachycaul: Having a thick trunk disproportionate to the rest of the plant

pachycladous: Having thick stems disproportionate to the rest of the plant

pad: A modified stem resembling a leaf as in the Opuntia species

palmate: 3 or more leaves radiating from a single point

pedicel: A flower stalk

peduncle: A stem or branch that holds a group of pedicels

perianth: The outer part of a flower surrounding the reproductive organs

petiole: The connecting stem from the leaf to a plant

pistil: The female reproductive organs of a flower including the stigma, style, and ovary which contains the seeds

polycarpic: An angiosperm that flowers and makes seeds multiple times in its life

propagation: The natural or artificial act of multiplying the offspring of plants

pruinose: Covered in a waxy white powder; farina

punctate: Pitted or dotted

-R-

radial: Structures radiating from a central point as in cactus spines

rhomboid: Diamond-shaped

rosette: The overlapping arrangement of leaves in a spiral to form a circular pattern that looks like a rose

rot: Decay in a plant caused by overwatering, bacteria, or fungi

-S-

scale: The shape of a bract or sepal

scale bug: A insect/pest that attaches itself to a plant; the disease name for a plant infested with scale insects

sepal: Non-fertile leaf-like structures covering a flower bud

sessile: Attached by the base directly to the stem rather than with a stalk or petiole; as with Echeveria leaves

serrate: Saw-toothed with teeth pointing towards leaf tip

simple: Having one leaf blade

sinuate: Margin with wavy indentations

sp.: Abbreviation for Species affinis, meaning a species is similar, related but not identical to the binomial name that follows; a single species; used when a species hasn't been identified

spp.: Abbreviation for two or more species

spatulate: Shaped like a spoon

stamen: The pollen forming organs of a flower

stigma: The part of a flower pistil that receives pollen, usually at the tip of the style

style: The stalk of a pistil between the stigma and the ovary

subsp.: Abbreviation for 'subspecies'

-T-

taproot: The main root of a plant

tepal: The modified leaves that make up a perianth

trichome: Hairlike growth on a plants surface; can be short or long

tubercle: A raised protuberance on the surface of a plant; the raised point where spines and flowers originate from cacti, as with many Mammillaria

-U-

umbel: Term for flowers that grow in a cluster from a common center, as with Hoya flowers

undulate: Shallow waves with wider spacing than sinuate margins

-V-

var.: Abbreviation for variety or verietas

variegated: Coloring that exhibits markings of another color, commonly white and green

vernalization: Period of cold temperatures that induce flowering

-W-

whorled: 3 or more leaves radiating from a single node

Sources

Biodiversity Heritage Library
Cactiguide.com
Curtis's Botanical Magazine
International Crassulaceae Network
JSTOR Global Plants Database
Llifle.com
Succulentguide.com

PLANT JOURNAL

"Mistress Mary always felt that however many years she lived she should never forget that first morning when her garden began to grow."

—Frances Hodgson Burnett, The Secret Garden, 1910

Monthly & Seasonal Notes

January:

February:

March:

April:

May:

June:

July:

August:

September:

October:

November:

December:

Winter:

Spring:

Summer:

Fall:

Date	Name

Notes

Date	Name

Date	Name

Date	Name

Notes

Index

(*Italicized page numbers refer to plant images*)

ANDREA AFRA, plant coach and life-long native of Houston, Texas, has helped millions of people around the world feel confident about caring for their succulents. Her educational website and videos at sucsforyou.com provide a much needed platform dedicated to succulent care with a focus on climate.

The Succulent Manual is based on the thousands of questions others have asked her over the years along with her own personal experience with growing more than 400 varieties of succulent plants. She spent her best childhood days helping her grandparents in the garden on their farm and continues to enjoy every moment tending and observing her plant collection.

Made in the USA
Middletown, DE
13 September 2023

38446980R00097